McGraw-Hill
Illustrative
Mathematics™
Course 2

Mc Graw Hill

mheducation.com/prek-12

Send all inquiries to:
McGraw-Hill Education
STEM Learning Solutions Center
8787 Orion Place
Columbus, OH 43240

ISBN: 978-0-07-689376-8
MHID: 0-07-689376-6

Illustrative Mathematics, Course 2
Student Edition, Volume 2

Printed in the United States of America.

8 9 10 11 12 LMN 28   27 26 25 24 23 22

Contents in Brief

Welcome to

McGraw-Hill
Illustrative
Mathematics

Learning math can be an exciting adventure.
That's why *Illustrative Mathematics* is designed
to provide opportunities for you to learn math
by doing math. By engaging in problem-solving
activities, you will make and test your own
conclusions and try out different strategies
as you learn math.

Let's get started!

Unit 1

Scale Drawings

dibrova/Shutterstock

Unit 2

Introducing Proportional Relationships

Collins93/Shutterstock

Unit 3

Measuring Circles

Blend Images/Image Source

Unit 4

Proportional Relationships and Percentages

Andrey Armyagov/123RF

Unit 5

Rational Number Arithmetic

Creative Travel Projects/Shutterstock

Unit 6

Expressions, Equations, and Inequalities

Daniel Dempster Photography/Alamy Stock Photo

Unit 7

Angles, Triangles, and Prisms

HDRExposed - Dave DiCello Photography/Flickr RF/Getty Images

Unit 8

Probability and Sampling

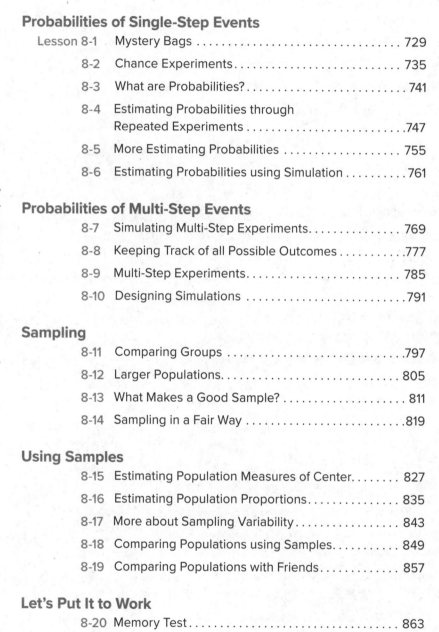

View Stock/View Stock RF/Getty Images

Unit 9

Putting It All Together

Corey Jenkins/Image Source

Expressions, Equations, and Inequalities

Daniel Dempster Photography/Alamy Stock Photo

Many state parks charge entrance fees by person and/or by vehicle. In this unit, you'll explore how expressions and equations can represent these fees.

Topics

- Representing Situations of the Form $px + q = r$ and $p(x + q) = r$
- Solving Equations of the Form $px + q = r$ and $p(x + q) = r$ and Problems that Lead to Those Equations
- Inequalities
- Writing Equivalent Expressions
- Let's Put It to Work

Unit 6
Expressions, Equations, and Inequalities

Lesson 6-1

Relationships between Quantities

NAME _____ DATE _____ PERIOD _____

Learning Goal Let's try to solve some new kinds of problems.

 ## Warm Up
1.1 Pricing Theater Popcorn

A movie theater sells popcorn in bags of different sizes. The table shows the volume of popcorn and the price of the bag.

Complete one column of the table with prices where popcorn is priced at a constant rate. That is, the amount of popcorn is proportional to the price of the bag. Then complete the other column with realistic example prices where the amount of popcorn and price of the bag are not in proportion.

Volume of Popcorn (ounces)	Price of Bag, Proportional ($)	Price of Bag, Not Proportional ($)
10	6	6
20		
35		
48		

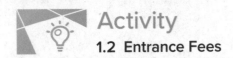

Activity

1.2 Entrance Fees

A state park charges an entrance fee based on the number of people in a vehicle. A car containing 2 people is charged $14, a car containing 4 people is charged $20, and a van containing 8 people is charged $32.

1. How much do you think a bus containing 30 people would be charged?

2. If a bus is charged $122, how many people do you think it contains?

3. What rule do you think the state park uses to decide the entrance fee for a vehicle?

Activity

1.3 Making Toast

A toaster has 4 slots for bread. Once the toaster is warmed up, it takes 35 seconds to make 4 slices of toast, 70 seconds to make 8 slices, and 105 seconds to make 10 slices.

1. How long do you think it will take to make 20 slices?

2. If someone makes as many slices of toast as possible in 4 minutes and 40 seconds, how many slices do you think they can make?

NAME _____ DATE _____ PERIOD _____

Are you ready for more?

What is the smallest number that has a remainder of 1, 2, and 3 when divided by 2, 3, and 4, respectively? Are there more numbers that have this property?

Summary
Relationships between Quantities

In much of our previous work that involved relationships between two quantities, we were often able to describe amounts as being so much more than another, or so many times as much as another. We wrote equations like $x + 3 = 8$ and $4x = 20$ and solved for unknown amounts.

In this unit, we will see situations where relationships between amounts involve more operations. For example, a pizza store might charge the amounts shown in the table for delivering pies.

Number of Pies	Total Cost in Dollars
1	13
2	23
3	33
5	53

We can see that each additional pie adds $10 to the total cost, and that each total includes a $3 additional cost, maybe representing a delivery fee. In this situation, 8 pies will cost $8 \cdot 10 + 3$ and a total cost of $63 means 6 pies were ordered.

In this unit, we will see many situations like this one, and will learn how to use diagrams and equations to answer questions about unknown amounts.

Practice
Relationships between Quantities

1. Lin and Tyler are drawing circles. Tyler's circle has twice the diameter of Lin's circle. Tyler thinks that his circle will have twice the area of Lin's circle as well. Do you agree with Tyler? (Lesson 3-7)

2. Jada and Priya are trying to solve the equation $\frac{2}{3} + x = 4$. (Lesson 5-15)

   - Jada says, "I think we should multiply each side by $\frac{3}{2}$ because that is the reciprocal of $\frac{2}{3}$."

   - Priya says, "I think we should add $-\frac{2}{3}$ to each side because that is the opposite of $\frac{2}{3}$."

   a. Which person's strategy should they use? Why?

   b. Write an equation that can be solved using the other person's strategy.

NAME _____ DATE _____ PERIOD _____

3. What are the missing operations? **(Lesson 5-13)**

   a. 48 ? (-8) = (-6)

   b. (-40) ? 8 =(-5)

   c. 12 ? (-2) = 14

   d. 18 ? (-12) = 6

   e. 18 ? (-20) = -2

   f. 22 ? (-0.5) = -11

4. In football, the team that has the ball has four chances to gain at least ten yards. If they don't gain at least ten yards, the other team gets the ball. Positive numbers represent a gain and negative numbers represent a loss. Select **all** of the sequences of four plays that result in the team getting to keep the ball. **(Lesson 5-14)**

   (A.) 8, -3, 4, 21

   (B.) 30, -7, -8, -12

   (C.) 2, 16, -5, -3

   (D.) 5, -2, 20, -1

   (E.) 20, -3, -13, 2

5. A sandwich store charges $20 to have 3 turkey subs delivered and $26 to have 4 delivered.

a. Is the relationship between the number of turkey subs delivered and the amount charged proportional? Explain how you know.

b. How much does the store charge for 1 additional turkey sub?

c. Describe a rule for determining how much the store charges based on the number of turkey subs delivered.

6. Which question cannot be answered by the solution to the equation $3x = 27$?

A. Elena read three times as many pages as Noah. She read 27 pages. How many pages did Noah read?

B. Lin has 27 stickers. She gives 3 stickers to each of her friends. With how many friends did Lin share her stickers?

C. Diego paid $27 to have 3 pizzas delivered and $35 to have 4 pizzas delivered. What is the price of one pizza?

D. The coach splits a team of 27 students into 3 groups to practice skills. How many students are in each group?

Lesson 6-2

Reasoning about Contexts with Tape Diagrams

NAME _____ DATE _____ PERIOD _____

Learning Goal Let's use tape diagrams to make sense of different kinds of stories.

 ## Warm Up
2.1 Notice and Wonder: Remembering Tape Diagrams

$$C$$

$a + b$	$a + b$	$a + b$	$a + b$

$$Z$$

x	x	x	x	y

1. What do you notice? What do you wonder?

2. What are some possible values for a, b, and c in the first diagram?
   For x, y, and z in the second diagram? How did you decide on those values?

Activity

2.2 Every Picture Tells a Story

Here are three stories with a diagram that represents it. With your group, decide who will go first. That person explains why the diagram represents the story. Work together to find any unknown amounts in the story. Then, switch roles for the second diagram and switch again for the third.

1. Mai made 50 flyers for five volunteers in her club to hang up around school. She gave 5 flyers to the first volunteer, 18 flyers to the second volunteer, and divided the remaining flyers equally among the three remaining volunteers.

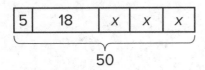

2. To thank her five volunteers, Mai gave each of them the same number of stickers. Then she gave them each two more stickers. Altogether, she gave them a total of 30 stickers.

$y+2$	$y+2$	$y+2$	$y+2$	$y+2$

30

3. Mai distributed another group of flyers equally among the five volunteers. Then she remembered that she needed some flyers to give to teachers, so she took 2 flyers from each volunteer. Then, the volunteers had a total of 40 flyers to hang up.

$w-2$	$w-2$	$w-2$	$w-2$	$w-2$

40

NAME _____ DATE _____ PERIOD _____

Activity

2.3 Every Story Needs a Picture

Here are three more stories. Draw a tape diagram to represent each story. Then describe how you would find any unknown amounts in the stories.

1. Noah and his sister are making gift bags for a birthday party. Noah puts 3 pencil erasers in each bag. His sister puts x stickers in each bag. After filling 4 bags, they have used a total of 44 items.

2. Noah's family also wants to blow up a total of 60 balloons for the party. Yesterday they blew up 24 balloons. Today they want to split the remaining balloons equally between four family members.

3. Noah's family bought some fruit bars to put in the gift bags. They bought one box each of four flavors: apple, strawberry, blueberry, and peach. The boxes all had the same number of bars. Noah wanted to taste the flavors and ate one bar from each box. There were 28 bars left for the gift bags.

Are you ready for more?

Design a tiling that uses a repeating pattern consisting of 2 kinds of shapes (e.g., 1 hexagon with 3 triangles forming a triangle). How many times did you repeat the pattern in your picture? How many individual shapes did you use?

Summary
Reasoning about Contexts with Tape Diagrams

Tape diagrams are useful for representing how quantities are related and can help us answer questions about a situation.

Suppose a school receives 46 copies of a popular book. The library takes 26 copies and the remainder are split evenly among 4 teachers. How many books does each teacher receive?

This situation involves 4 equal parts and one other part.

- We can represent the situation with a rectangle labeled 26 (books given to the library) along with 4 equal-sized parts (books split among 4 teachers).

- We label the total, 46, to show how many the rectangle represents in all.

- We use a letter to show the unknown amount, which represents the number of books each teacher receives.

- Using the same letter, x, means that the same number is represented four times.

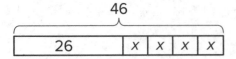

Some situations have parts that are all equal, but each part has been increased from an original amount:

A company manufactures a special type of sensor, and packs them in boxes of 4 for shipment. Then a new design increases the weight of each sensor by 9 grams. The new package of 4 sensors weighs 76 grams. How much did each sensor weigh originally?

We can describe this situation with a rectangle representing a total of 76 split into 4 equal parts. Each part shows that the new weight, $x + 9$, is 9 more than the original weight, x.

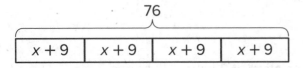

NAME _____ DATE _____ PERIOD _____

Practice
Reasoning about Contexts with Tape Diagrams

1. The table shows the number of apples and the total weight of the apples. (Lesson 3-1)

Number of Apples	Weight of Apples (grams)
2	511
5	1200
8	2016

Estimate the weight of 6 apples.

2. Select **all** stories that the tape diagram can represent.

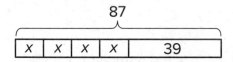

A. There are 87 children and 39 adults at a show. The seating in the theater is split into 4 equal sections.

B. There are 87 first graders in after-care. After 39 students are picked up, the teacher put the remaining students into 4 groups for an activity.

C. Lin buys a pack of 87 pencils. She gives 39 to her teacher and shared the remaining pencils between herself and 3 friends.

D. Andre buys 4 packs of paper clips with 39 paper clips in each. Then he gives 87 paper clips to his teacher.

E. Diego's family spends $87 on 4 tickets to the fair and a $39 dinner.

3. Andre wants to save $40 to buy a gift for his dad. Andre's neighbor will pay him weekly to mow the lawn, but Andre always gives a $2 donation to the food bank in weeks when he earns money. Andre calculates that it will take him 5 weeks to earn the money for his dad's gift. He draws a tape diagram to represent the situation.

$$
\underbrace{\boxed{x-2}\,\boxed{x-2}\,\boxed{x-2}\,\boxed{x-2}\,\boxed{x-2}}_{40}
$$

a. Explain how the parts of the tape diagram represent the story.

b. How much does Andre's neighbor pay him each week to mow the lawn?

4. Without evaluating each expression, determine which value is the greatest. Explain how you know. **(Lesson 5-13)**

$$7\tfrac{5}{6} - 9\tfrac{3}{4}$$

$$\left(-7\tfrac{5}{6}\right) + \left(-9\tfrac{3}{4}\right)$$

$$\left(-7\tfrac{5}{6}\right) \cdot \left(9\tfrac{3}{4}\right)$$

$$\left(-7\tfrac{5}{6}\right) \div \left(-9\tfrac{3}{4}\right)$$

5. Solve each equation. **(Lesson 5-15)**

a. $(8.5) \cdot (-3) = a$

b. $(-7) + b = (-11)$

c. $c - (-3) = 15$

d. $d \cdot (-4) = 32$

Lesson 6-3

Reasoning about Equations with Tape Diagrams

NAME _____ DATE _____ PERIOD _____

Learning Goal Let's see how equations can describe tape diagrams.

Warm Up

3.1 Find Equivalent Expressions

Select **all** the expressions that are equivalent to $7(2 - 3n)$. Explain how you know each expression you select is equivalent.

A. $9 - 10n$

B. $14 - 3n$

C. $14 - 21n$

D. $(2 - 3n) \cdot 7$

E. $7 \cdot 2 \cdot (-3n)$

Activity

3.2 Matching Equations to Tape Diagrams

1. Match each equation to one of the tape diagrams. Be prepared to explain how the equation matches the diagram.

Equations

- $2x + 5 = 19$
- $2 + 5x = 19$
- $2(x + 5) = 19$
- $5(x + 2) = 19$
- $19 = 5 + 2x$
- $(x + 5) \cdot 2 = 19$
- $19 = (x + 2) \cdot 5$
- $19 \div 2 = x + 5$
- $19 - 2 = 5x$

Diagrams

A
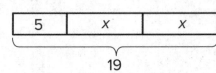

B
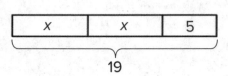

C
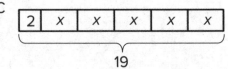

D
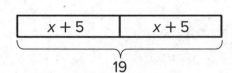

E
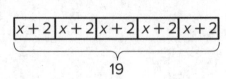

2. Sort the equations into categories of your choosing. Explain the criteria for each category.

NAME _____ DATE _____ PERIOD _____

Activity

3.3 Drawing Tape Diagrams to Represent Equations

1. Draw a tape diagram to match each equation.

   - $114 = 3x + 18$

   - $114 = 3(y + 18)$

2. Use any method to find values for x and y that make the equations true.

Are you ready for more?

To make a Koch snowflake:

- Start with an equilateral triangle that has side lengths of 1. This is Step 1.

- Replace the middle third of each line segment with a small equilateral triangle with the middle third of the segment forming the base. This is Step 2.

- Do the same to each of the line segments. This is Step 3.

- Keep repeating this process.

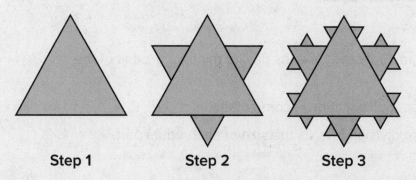

Step 1   Step 2   Step 3

1. What is the perimeter after Step 2? Step 3?

2. What happens to the perimeter, or the length of line traced along the outside of the figure, as the process continues?

 ## Summary
Reasoning about Equations with Tape Diagrams

We have seen how tape diagrams represent relationships between quantities. Because of the meaning and properties of addition and multiplication, more than one equation can often be used to represent a single tape diagram.

Let's take a look at two tape diagrams.

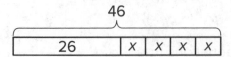

We can describe this diagram with several different equations. Here are some of them.

- $26 + 4x = 46$, because the parts add up to the whole.

- $4x + 26 = 46$, because addition is commutative.

- $46 = 4x + 26$, because if two quantities are equal, it doesn't matter how we arrange them around the equal sign.

- $4x = 46 - 26$, because one part (the part made up of 4 x's) is the difference between the whole and the other part.

76			
$x + 9$	$x + 9$	$x + 9$	$x + 9$

For this diagram:

- $4(x + 9) = 76$, because multiplication means having multiple groups of the same size.

- $(x + 9) \cdot 4 = 76$, because multiplication is commutative.

- $76 \div 4 = x + 9$, because division tells us the size of each equal part.

Glossary

equivalent expressions

NAME _____ DATE _____ PERIOD _____

Practice
Reasoning about Equations with Tape Diagrams

1. Solve each equation mentally. (Lesson 5-15)

   a. $2x = 10$

   b. $-3x = 21$

   c. $\frac{1}{3}x = 6$

   d. $-\frac{1}{2}x = -7$

2. Complete the magic squares so that the sum of each row, each column, and each diagonal in a grid are all equal. (Lesson 5-3)

0	7	2
	3	

1		
	3	-2
		5

4	2	0
-1		

3. Draw a tape diagram to match each equation.

   a. $5(x + 1) = 20$

   b. $5x + 1 = 20$

4. Select **all** the equations that match the tape diagram.

$$35$$

8	x	x	x	x	x	x

(A.) $35 = 8 + x + x + x + x + x + x$

(B.) $35 = 8 + 6x$

(C.) $6 + 8x = 35$

(D.) $6x + 8 = 35$

(E.) $6x + 8x = 35x$

(F.) $35 - 8 = 6x$

5. Each car is traveling at a constant speed. Find the number of miles each car travels in 1 hour at the given rate. **(Lesson 4-2)**

a. 135 miles in 3 hours

b. 22 miles in $\frac{1}{2}$ hour

c. 7.5 miles in $\frac{1}{4}$ hour

d. $\frac{100}{3}$ miles in $\frac{2}{3}$ hour

e. $97\frac{1}{2}$ miles in $\frac{3}{2}$ hour

Lesson 6-4

Reasoning about Equations and Tape Diagrams (Part 1)

NAME _____ DATE _____ PERIOD _____

Learning Goal Let's see how tape diagrams can help us answer questions about unknown amounts in stories.

 ## Warm Up
4.1 Algebra Talk: Seeing Structure

Find a solution to each equation without writing anything down.

1. $x + 1 = 5$ **2.** $2(x + 1) = 10$

3. $3(x + 1) = 15$ **4.** $500 = 100(x + 1)$

 ## Activity
4.2 Situations and Diagrams

Draw a tape diagram to represent each situation. For some of the situations, you need to decide what to represent with a variable.

1. Diego has 7 packs of markers. Each pack has x markers in it. After Lin gives him 9 more markers, he has a total of 30 markers.

2. Elena is cutting a 30-foot piece of ribbon for a craft project. She cuts off 7 feet, and then cuts the remaining piece into 9 equal lengths of x feet each.

3. A construction manager weighs a bundle of 9 identical bricks and a 7-pound concrete block. The bundle weighs 30 pounds.

4. A skating rink charges a group rate of $9 plus a fee to rent each pair of skates. A family rents 7 pairs of skates and pays a total of $30.

5. Andre bakes 9 pans of brownies. He donates 7 pans to the school bake sale and keeps the rest to divide equally among his class of 30 students.

 Activity

4.3 Situations, Diagrams, and Equations

Each situation in the previous activity is represented by one of the equations.

- $7x + 9 = 30$
- $30 = 9x + 7$
- $30x + 7 = 9$

1. Match each situation to an equation.

2. Find the solution to each equation. Use your diagrams to help you reason.

3. What does each solution tell you about its situation?

Are you ready for more?

While in New York City, is it a better deal for a group of friends to take a taxi or the subway to get from the Empire State Building to the Metropolitan Museum of Art? Explain your reasoning.

NAME _____ DATE _____ PERIOD _____

Summary

Reasoning about Equations and Tape Diagrams (Part 1)

Many situations can be represented by equations. Writing an equation to represent a situation can help us express how quantities in the situation are related to each other, and can help us reason about unknown quantities whose value we want to know. Here are three situations.

- An architect is drafting plans for a new supermarket. There will be a space 144 inches long for rows of nested shopping carts. The first cart is 34 inches long and each nested cart adds another 10 inches. The architect wants to know how many shopping carts will fit in each row.

- A bakery buys a large bag of sugar that has 34 cups. They use 10 cups to make some cookies. Then they use the rest of the bag to make 144 giant muffins. Their customers want to know how much sugar is in each muffin.

- Kiran is trying to save $144 to buy a new guitar. He has $34 and is going to save $10 a week from money he earns mowing lawns. He wants to know how many weeks it will take him to have enough money to buy the guitar.

We see the same three numbers in the situations: 10, 34, and 144. How could we represent each situation with an equation?

In the first situation, there is one shopping cart with length 34 and then an unknown number of carts with length 10. Similarly, Kiran has 34 dollars saved and then will save 10 each week for an unknown number of weeks. Both situations have one part of 34 and then equal parts of size 10 that all add together to 144. Their equation is $34 + 10x = 144$.

Since it takes 11 groups of 10 to get from 34 to 144, the value of x in these two situations is $(144 - 34) \div 10$ or 11. There will be 11 nested shopping carts in each row, and it will take Kiran 11 weeks to raise the money for the guitar.

In the bakery situation, there is one part of 10 and then 144 equal parts of unknown size that all add together to 34. The equation is $10 + 144x = 34$.

Since 24 is needed to get from 10 to 34, the value of x is $(34 - 10) \div 144$ or $\frac{1}{6}$. There is $\frac{1}{6}$ cup of sugar in each giant muffin.

Practice

Reasoning about Equations and Tape Diagrams (Part 1)

1. Draw a square with side length 7 cm.
   (Lesson 3-1)

   a. Predict the perimeter and the length of the diagonal of the square.

   b. Measure the perimeter and the length of the diagonal of the square.

   c. Describe how close the predictions and measurements are.

2. Find the products. (Lesson 5-9)

   a. $(100) \cdot (-0.09)$

   b. $(-7) \cdot (-1.1)$

   c. $(-7.3) \cdot (5)$

   d. $(-0.2) \cdot (-0.3)$

NAME _____ DATE _____ PERIOD _____

3. Here are three stories.

- A family buys 6 tickets to a show. They also pay a $3 parking fee. They spend $27 to see the show.

- Diego has 27 ounces of juice. He pours equal amounts for each of his 3 friends and has 6 ounces left for himself.

- Jada works for 6 hours preparing for the art fair. She spends 3 hours on a sculpture and then paints 27 picture frames.

Here are three equations.

- $3x + 6 = 27$

- $6x + 3 = 27$

- $27x + 3 = 6$

a. Decide which equation represents each story. What does x represent in each equation?

b. Find the solution to each equation. Explain or show your reasoning.

c. What does each solution tell you about its situation?

4. Here is a diagram and its corresponding equation. Find the solution to the equation and explain your reasoning.

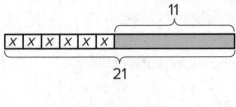

$$6x + 11 = 21$$

5. a. Plot these points on the coordinate plane: (Lesson 5-7)

$A = (3,2), B = (7.5,2), C = (7.5,-2.5), D = (3,-2)$

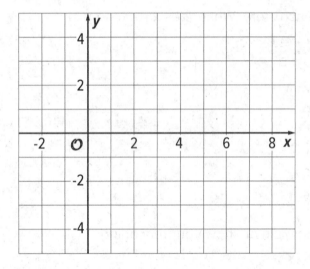

b. What is the vertical difference between D and A?

c. Write an expression that represents the vertical distance between B and C.

Lesson 6-5

Reasoning about Equations and Tape Diagrams (Part 2)

NAME _____ DATE _____ PERIOD _____

Learning Goal Let's use tape diagrams to help answer questions about situations where the equation has parentheses.

 ## Warm Up
5.1 Algebra Talk: Seeing Structure

Solve each equation mentally.

1. $x - 1 = 5$ **2.** $2(x - 1) = 10$

3. $3(x - 1) = 15$ **4.** $500 = 100(x - 1)$

 ## Activity
5.2 More Situations and Diagrams

Draw a tape diagram to represent each situation. For some of the situations, you need to decide what to represent with a variable.

1. Each of 5 gift bags contains x pencils. Tyler adds 3 more pencils to each bag. Altogether, the gift bags contain 20 pencils.

2. Noah drew an equilateral triangle with sides of length 5 inches. He wants to increase the length of each side by x inches so the triangle is still equilateral and has a perimeter of 20 inches.

3. An art class charges each student $3 to attend plus a fee for supplies. Today, $20 was collected for the 5 students attending the class.

4. Elena ran 20 miles this week, which was three times as far as Clare ran this week. Clare ran 5 more miles this week than she did last week.

 Activity

5.3 More Situations, Diagrams, and Equations

Each situation in the previous activity is represented by one of the equations.

- $(x + 3) \cdot 5 = 20$
- $3(x + 5) = 20$

1. Match each situation to an equation.

2. Find the solution to each equation. Use your diagrams to help you reason.

3. What does each solution tell you about its situation?

NAME _____ DATE _____ PERIOD _____

Are you ready for more?

Han, his sister, his dad, and his grandmother step onto a crowded bus with only 3 open seats for a 42-minute ride. They decide Han's grandmother should sit for the entire ride. Han, his sister, and his dad take turns sitting in the remaining two seats, and Han's dad sits 1.5 times as long as both Han and his sister. How many minutes did each one spend sitting?

Summary

Reasoning about Equations and Tape Diagrams (Part 2)

Equations with parentheses can represent a variety of situations.

- Lin volunteers at a hospital and is preparing toy baskets for children who are patients. She adds 2 items to each basket, after which the supervisor's list shows that 140 toys have been packed into a group of 10 baskets. Lin wants to know how many toys were in each basket before she added the items.

- A large store has the same number of workers on each of 2 teams to handle different shifts. They decide to add 10 workers to each team, bringing the total number of workers to 140. An executive at the company that runs this chain of stores wants to know how many employees were in each team before the increase.

Each bag in the first story has an unknown number of toys, x, that is increased by 2. Then ten groups of $x + 2$ give a total of 140 toys. An equation representing this situation is $10(x + 2) = 140$. Since 10 times a number is 140, that number is 14, which is the total number of items in each bag. Before Lin added the 2 items there were $14 - 2$ or 12 toys in each bag.

The executive in the second story knows that the size of each team of y employees has been increased by 10. There are now 2 teams of $y + 10$ each. An equation representing this situation is $2(y + 10) = 140$. Since 2 times an amount is 140, that amount is 70, which is the new size of each team. The value of y is $70 - 10$ or 60. There were 60 employees on each team before the increase.

Reasoning about Equations and Tape Diagrams (Part 2)

1. Here are some prices customers paid for different items at a farmer's market. Find the cost for 1 pound of each item. (Lesson 4-2)

   a. $5 for 4 pounds of apples

   b. $3.50 for $\frac{1}{2}$ pound of cheese

   c. $8.25 for $1\frac{1}{2}$ pounds of coffee beans

   d. $6.75 for $\frac{3}{4}$ pounds of fudge

   e. $5.50 for $6\frac{1}{4}$ pound pumpkin

2. Find the products. (Lesson 5-9)

   a. $\frac{2}{3} \cdot \left(\frac{-4}{5}\right)$

   b. $\left(\frac{-5}{7}\right) \cdot \left(\frac{7}{5}\right)$

   c. $\left(\frac{-2}{39}\right) \cdot 39$

   d. $\left(\frac{2}{5}\right) \cdot \left(\frac{-3}{4}\right)$

NAME _____ DATE _____ PERIOD _____

3. Here are two stories.

   • A family buys 6 tickets to a show. They also *each* spend $3 on a snack. They spend $24 on the show.

   • Diego has 24 ounces of juice. He pours equal amounts for each of his 3 friends, and then adds 6 more ounces for each.

   Here are two equations.

   • $3(x + 6) = 24$

   • $6(x + 3) = 24$

   a. Which equation represents which story?

   b. What does x represent in each equation?

   c. Find the solution to each equation. Explain or show your reasoning.

   d. What does each solution tell you about its situation?

4. Here is a diagram and its corresponding equation. Find the solution to the equation and explain your reasoning.

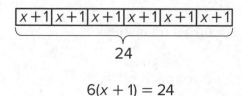

$$6(x + 1) = 24$$

5. Below is a set of data about temperatures. The *range* of a set of data is the distance between the lowest and highest value in the set. What is the range of these temperatures? (Lesson 5-7)

9°C, -3°C, 22°C, -5°C, 11°C, 15°C

6. A store is having a 25% off sale on all shirts. Show two different ways to calculate the sale price for a shirt that normally costs $24. (Lesson 4-11)

Lesson 6-6

Distinguishing between Two Types of Situations

NAME _____ DATE _____ PERIOD _____

Learning Goal Let's think about equations with and without parentheses and the kinds of situations they describe.

Warm Up
6.1 Which One Doesn't Belong: Seeing Structure

Which equation doesn't belong?

1. $4(x + 3) = 9$

2. $4 \cdot x + 12 = 9$

3. $4 + 3x = 9$

4. $9 = 12 + 4x$

Activity
6.2 Card Sort: Categories of Equations

Your teacher will give you a set of cards that show equations. Sort the cards into 2 categories of your choosing. Be prepared to explain the meaning of your categories. Then, sort the cards into 2 categories in a different way.

Be prepared to explain the meaning of your new categories.

Diagram A

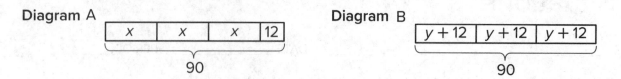

Diagram B

Story 1: Lin had 90 flyers to hang up around the school. She gave 12 flyers to each of three volunteers. Then she took the remaining flyers and divided them up equally among the three volunteers.

Story 2: Lin had 90 flyers to hang up around the school. After giving the same number of flyers to each of three volunteers, she had 12 left to hang up by herself.

1. Which diagram goes with which story? Be prepared to explain your reasoning.

2. In each diagram, what part of the story does the variable represent?

3. Write an equation corresponding to each story. If you get stuck, use the diagram.

4. Find the value of the variable in each story.

NAME _____ DATE _____ PERIOD _____

Are you ready for more?

A tutor is starting a business. In the first year, they start with 5 clients and charge $10 per week for an hour of tutoring with each client. For each year following, they double the number of clients and the number of hours each week. Each new client will be charged 150% of the charges of the clients from the previous year.

1. Organize the weekly earnings for each year in a table.

2. Assuming a full-time week is 40 hours per week, how many years will it take to reach full time and how many new clients will be taken on that year?

3. After reaching full time, what is the tutor's annual salary if they take 2 weeks of vacation?

4. Is there another business model you'd recommend for the tutor? Explain your reasoning.

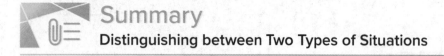

In this unit, we encounter two main types of situations that can be represented with an equation. Here is an example of each type.

- After adding 8 students to each of 6 same-sized teams, there were 72 students altogether.

- After adding an 8-pound box of tennis rackets to a crate with 6 identical boxes of ping pong paddles, the crate weighed 72 pounds.

The first situation has all equal parts, since additions are made to *each* team. An equation that represents this situation is $6(x + 8) = 72$, where x represents the original number of students on each team. Eight students were added to each group, there are 6 groups, and there are a total of 72 students.

In the second situation, there are 6 equal parts added to one other part. An equation that represents this situation is $6x + 8 = 72$, where x represents the weight of a box of ping pong paddles, there are 6 boxes of ping pong paddles, there is an additional box that weighs 8 pounds, and the crate weighs 72 pounds altogether.

- In the first situation, there were 6 equal groups, and 8 students added to each group. $6(x + 8) = 72$.

- In the second situation, there were 6 equal groups, but 8 more pounds in addition to that. $6x + 8 = 72$.

NAME _____ DATE _____ PERIOD _____

Practice
Distinguishing between Two Types of Situations

1. A school ordered 3 large boxes of board markers. After giving 15 markers to each of 3 teachers, there were 90 markers left. The diagram represents the situation. How many markers were originally in each box? (Lesson 6-2)

$$\overbrace{\boxed{x - 15 \,|\, x - 15 \,|\, x - 15}}^{90}$$

2. The diagram can be represented by the equation $25 = 2 + 6x$. Explain where you can see the 6 in the diagram. (Lesson 6-3)

$$\overbrace{\boxed{2 \,|\, x \,|\, x \,|\, x \,|\, x \,|\, x \,|\, x}}^{25}$$

3. Match each equation to a story. (Two of the stories match the same equation.)

Equations

a. $3(x + 5) = 17$

c. $5(x + 3) = 17$

b. $3x + 5 = 17$

d. $5x + 3 = 17$

Stories

i. Jada's teacher fills a travel bag with 5 copies of a textbook. The weight of the bag and books is 17 pounds. The empty travel bag weighs 3 pounds. How much does each book weigh?

ii. A piece of scenery for the school play is in the shape of a 5-foot-long rectangle. The designer decides to increase the length. There will be 3 identical rectangles with a total length of 17 feet. By how much did the designer increase the length of each rectangle?

iii. Elena spends $17 and buys a $3 book and a bookmark for each of her 5 cousins. How much does each bookmark cost?

iv. Noah packs up bags at the food pantry to deliver to families. He packs 5 bags that weigh a total of 17 pounds. Each bag contains 3 pounds of groceries and a packet of papers with health-related information. How much does each packet of papers weigh?

v. Andre has 3 times as many pencils as Noah and 5 pens. He has 17 pens and pencils all together. How many pencils does Noah have?

4. Elena walked 20 minutes more than Lin. Jada walked twice as long as Elena. Jada walked for 90 minutes. The equation $2(x + 20) = 90$ describes this situation. Match each expression with the statement in the story the expression represents.

Expressions

a. x

b. $x + 20$

c. $2(x + 20)$

d. 90

Statements

i. the number of minutes that Jada walked

ii. the number of minutes that Elena walked

iii. the number of minutes that Lin walked

Lesson 6-7

Reasoning about Solving Equations (Part 1)

NAME _____ DATE _____ PERIOD _____

Learning Goal Let's see how a balanced hanger is like an equation and how moving its weights is like solving the equation.

 Warm Up

7.1 Hanger Diagrams

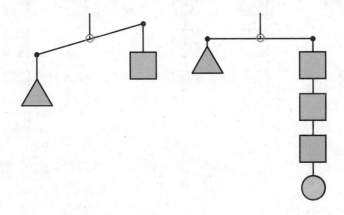

In the two diagrams, all the triangles weigh the same and all the squares weigh the same.

For each diagram, come up with . . .

1. One thing that *must* be true

2. One thing that *could* be true

3. One thing that *cannot possibly* be true

Activity

7.2 Hanger and Equation Matching

On each balanced hanger, figures with the same letter have the same weight.

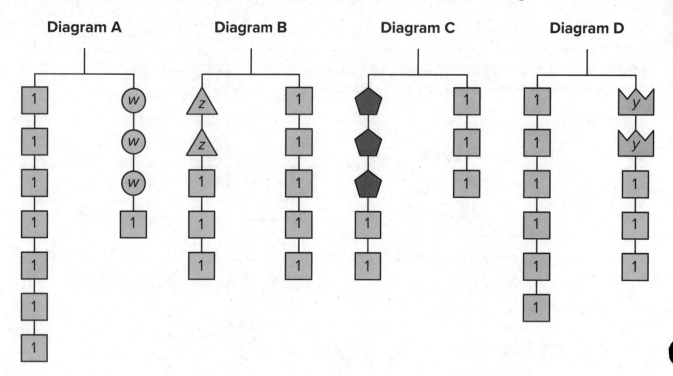

Diagram A   **Diagram B**   **Diagram C**   **Diagram D**

1. Match each hanger to an equation. Complete the equation by writing *x, y, z,* or *w* in the empty box.

   - $2\boxed{} + 3 = 5$
   - $3\boxed{} + 2 = 3$
   - $6 = 2\boxed{} + 3$
   - $7 = 3\boxed{} + 1$

2. Find the solution to each equation. Use the hanger to explain what the solution means.

NAME _____ DATE _____ PERIOD _____

 Activity
7.3 Use Hangers to Understand Equation Solving

Here are some balanced hangers where each piece is labeled with its weight.

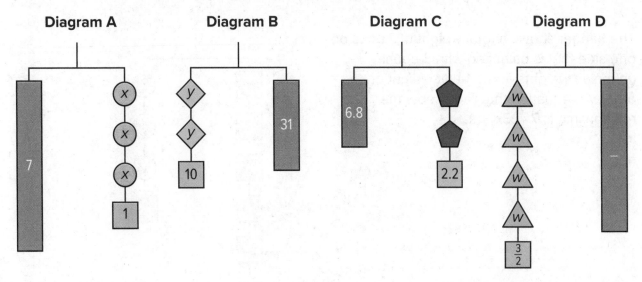

Diagram A   **Diagram B**   **Diagram C**   **Diagram D**

For each diagram:

1. Write an equation.

2. Explain how to figure out the weight of a piece labeled with a letter by reasoning about the diagram.

3. Explain how to figure out the weight of a piece labeled with a letter by reasoning about the equation.

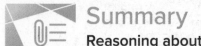

Summary
Reasoning about Solving Equations (Part 1)

In this lesson, we worked with two ways to show that two amounts are equal: a balanced hanger and an equation. We can use a balanced hanger to think about steps to finding an unknown amount in an associated equation.

The hanger shows a total weight of 7 units on one side that is balanced with 3 equal, unknown weights and a 1-unit weight on the other. An equation that represents the relationship is $7 = 3x + 1$.

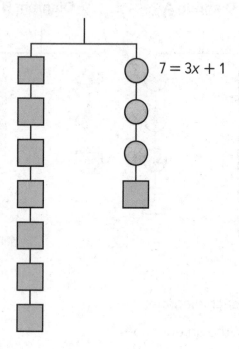

$$7 = 3x + 1$$

We can remove a weight of 1 unit from each side and the hanger will stay balanced. This is the same as subtracting 1 from each side of the equation.

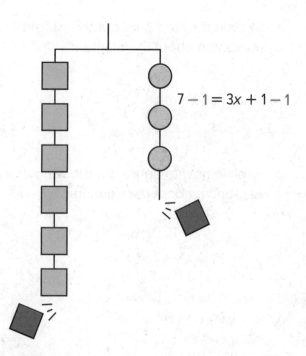

$$7 - 1 = 3x + 1 - 1$$

NAME _____ DATE _____ PERIOD _____

An equation for the new balanced hanger is $6 = 3x$.

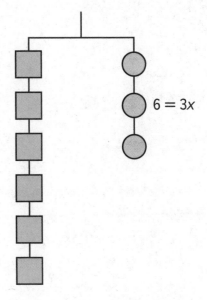

So the hanger will balance with $\frac{1}{3}$ of the weight on each side: $\frac{1}{3} \cdot 6 = \frac{1}{3} \cdot 3x$.

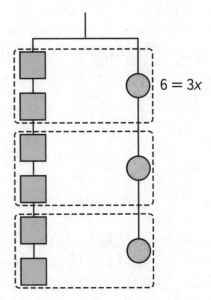

The two sides of the hanger balance with these weights: 6 1-unit weights on one side and 3 weights of unknown size on the other side.

Here is a concise way to write the steps above:

$7 = 3x + 1$

$6 = 3x$ after subtracting 1 from each side

$2 = x$ after multiplying each side by $\frac{1}{3}$

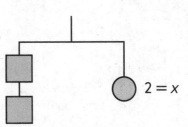

1. There is a proportional relationship between the volume of a sample of helium in liters and the mass of that sample in grams. If the mass of a sample is 5 grams, its volume is 28 liters. (5, 28) is shown on the graph below. (Lesson 2-11)

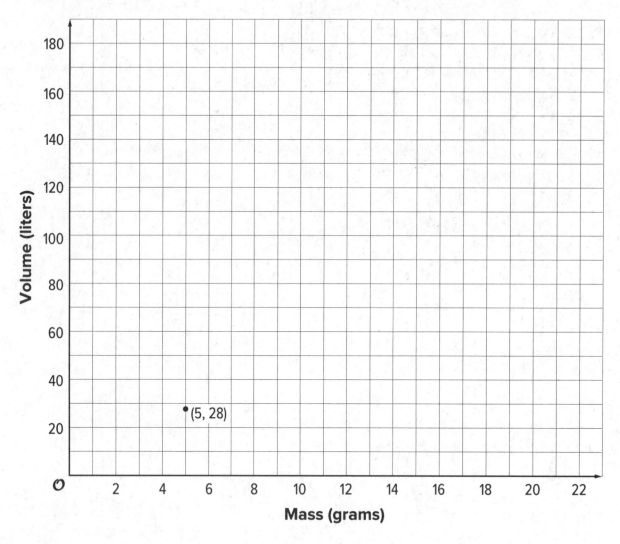

a. What is the constant of proportionality in this relationship?

NAME _____ DATE _____ PERIOD _____

b. In this situation, what is the meaning of the number you found in part a?

c. Add at least three more points to the graph, and label with their coordinates.

d. Write an equation that shows the relationship between the mass of a sample of helium and its volume. Use m for mass and v for volume.

2. Explain how the parts of the balanced hanger compare to the parts of the equation.

$7 = 2x + 3$

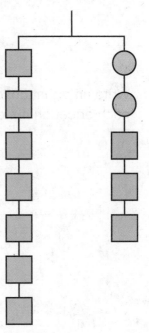

3. Here is a hanger.

a. Write an equation to represent the hanger.

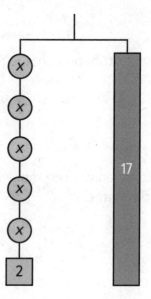

b. Draw more hangers to show each step you would take to find x. Explain your reasoning.

c. Write an equation to describe each hanger you drew. Describe how each equation matches its hanger.

Lesson 6-8

Reasoning about Solving Equations (Part 2)

NAME _____ DATE _____ PERIOD _____

Learning Goal Let's use hangers to understand two different ways of solving equations with parentheses.

Warm Up
8.1 Equivalent to 2(x + 3)

Select **all** the expressions equivalent to $2(x + 3)$.

(A.) $2 \cdot (x + 3)$ (D.) $2 \cdot x + 3$

(B.) $(x + 3)2$ (E.) $(2 \cdot x) + 3$

(C.) $2 \cdot x + 2 \cdot 3$ (F.) $(2 + x)3$

Activity
8.2 Either Or

1. Explain why either of these equations could represent this hanger.

   $14 = 2(x + 3)$ or $14 = 2x + 6$

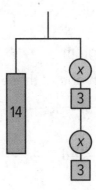

2. Find the weight of one circle. Be prepared to explain your reasoning.

Activity

8.3 Use Hangers to Understand Equation Solving, Again

Here are some balanced hangers. Each piece is labeled with its weight.

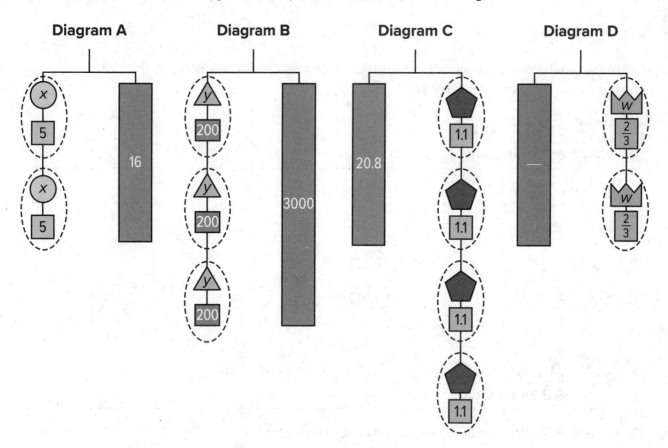

Diagram A   **Diagram B**   **Diagram C**   **Diagram D**

For each diagram:

  a. Assign one of these equations to each hanger:

$$2(x + 5) = 16 \qquad\qquad\qquad 20.8 = 4(z + 1.1)$$

$$3(y + 200) = 3{,}000 \qquad\qquad \frac{20}{3} = 2\left(w + \frac{2}{3}\right)$$

  b. Explain how to figure out the weight of a piece labeled with a letter by reasoning about the diagram.

  c. Explain how to figure out the weight of a piece labeled with a letter by reasoning about the equation.

NAME _____ DATE _____ PERIOD _____

Summary
Reasoning about Solving Equations (Part 2)

The balanced hanger shows 3 equal, unknown weights and 3 2-unit weights on the left and an 18-unit weight on the right.

There are 3 unknown weights plus 6 units of weight on the left. We could represent this balanced hanger with an equation and solve the equation the same way we did before.

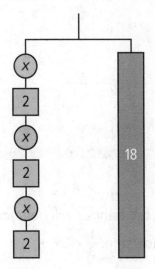

$$3x + 6 = 18$$
$$3x = 12$$
$$x = 4$$

Since there are 3 groups of $x + 2$ on the left, we could represent this hanger with a different equation: $3(x + 2) = 18$.

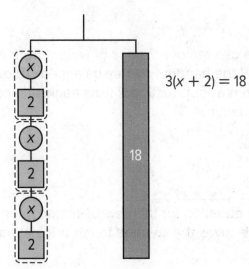

$$3(x + 2) = 18$$

The two sides of the hanger balance with these weights: 3 groups of $x + 2$ on one side, and 18, or 3 groups of 6, on the other side.

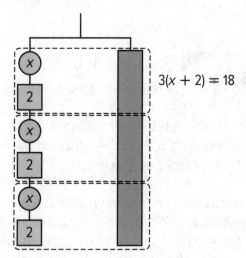

$$3(x + 2) = 18$$

The two sides of the hanger will balance with $\frac{1}{3}$ of the weight on each side: $\frac{1}{3} \cdot (3x + 2) = \frac{1}{3} \cdot 18$.

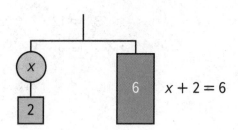

$$x + 2 = 6$$

We can remove 2 units of weight from each side, and the hanger will stay balanced. This is the same as subtracting 2 from each side of the equation.

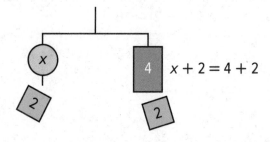

$$x + 2 = 4 + 2$$

An equation for the new balanced hanger is $x = 4$. This gives the solution to the original equation.

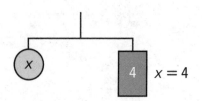

$$x = 4$$

Here is a concise way to write the steps above:

$3(x + 2) = 18$

$\quad x + 2 = 6 \qquad$ after multiplying each side by $\frac{1}{3}$

$\qquad x = 4 \qquad$ after subtracting 2 from each side

NAME _____ DATE _____ PERIOD _____

Practice
Reasoning about Solving Equations (Part 2)

1. Here is a hanger.

   a. Write an equation to represent the hanger.

   b. Solve the equation by reasoning about the
   equation or the hanger. Explain your reasoning.

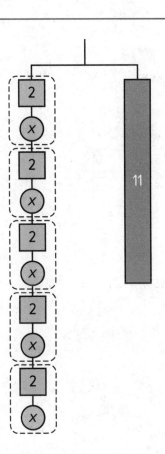

2. Explain how each part of the equation $9 = 3(x + 2)$ is represented in the hanger.

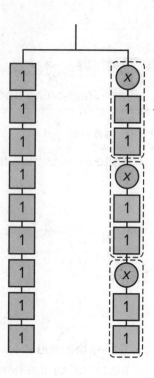

   a. x

   b. 9

   c. 3

   d. $x + 2$

   e. $3(x + 2)$

   f. the equal sign

NAME _____ DATE _____ PERIOD _____

3. Select the word from the following list that best describes each
   situation. (Lesson 4-11)

   Tax   Discount   Tip or gratuity

   Commission   Markup   Interest

   a. You deposit money in a savings account, and every year
      the amount of money in the account increases by 2.5%.

   b. For every car sold, a car salesman is paid 6% of the
      car's price.

   c. Someone who eats at a restaurant pays an extra 20% of the
      food price. This extra money is kept by the person who
      served the food.

   d. An antique furniture store pays $200 for a chair, adds 50%
      of that amount, and sells the chair for $300.

   e. The normal price of a mattress is $600, but it is on sale for
      10% off.

   f. For any item you purchase in Texas, you pay an additional
      6.25% of the item's price to the state government.

4. Clare drew this diagram to match the equation $2x + 16 = 50$, but she got the wrong solution as a result of using this diagram. **(Lesson 6-3)**

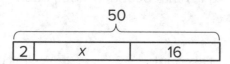

a. What value for x can be found using the diagram?

b. Show how to fix Clare's diagram to correctly match the equation.

c. Use the new diagram to find a correct value for x.

d. Explain the mistake Clare made when she drew her diagram.

Lesson 6-9

Dealing with Negative Numbers

NAME _____ DATE _____ PERIOD _____

Learning Goal Let's show that doing the same to each side works for negative numbers too.

 ## Warm Up
9.1 Which One Doesn't Belong: Rational Number Arithmetic

Which equation doesn't belong?

A. $15 = \text{-}5 \cdot \text{-}3$ **C.** $4 - \text{-}2 = 6$

B. $2 + \text{-}5 = \text{-}3$ **D.** $\text{-}3 \cdot \text{-}4 = \text{-}12$

 ## Activity
9.2 Old and New Ways to Solve

Solve each equation. Be prepared to explain your reasoning.

1. $x + 6 = 4$

2. $x - \text{-}4 = \text{-}6$

3. $2(x - 1) = \text{-}200$

4. $2x + \text{-}3 = \text{-}23$

Here are some equations that all have the same solution.

$$x = -6$$
$$x - 3 = -9$$
$$-9 = x - 3$$
$$900 = -100(x - 3)$$
$$900 = (x - 3) \cdot (-100)$$
$$900 = -100x + 300$$

1. Explain how you know that each equation has the same solution as the previous equation. Pause for discussion before moving to the next question.

2. Keep your work secret from your partner. Start with the equation $-5 = x$. Do the same thing to each side at least three times to create an equation that has the same solution as the starting equation. Write the equation you ended up with on a slip of paper, and trade equations with your partner.

3. See if you can figure out what steps they used to transform $-5 = x$ into their equation. When you think you know, check with them to see if you are right.

NAME _____ DATE _____ PERIOD _____

Summary
Dealing with Negative Numbers

When we want to find a solution to an equation, sometimes we just think about what value in place of the variable would make the equation true. Sometimes we perform the same operation on each side (for example, subtract the same amount from each side). The balanced hangers helped us to understand that doing the same to each side of an equation keeps the equation true.

Since negative numbers are just numbers, then doing the same thing to each side of an equation works for negative numbers as well. Here are some examples of equations that have negative numbers and steps you could take to solve them.

Example:

$$2(x - 5) = -6$$

$$\frac{1}{2} \cdot 2(x - 5) = \frac{1}{2} \cdot (-6) \qquad \text{Multiply each side by } \frac{1}{2}.$$

$$x - 5 = -3$$

$$x - 5 + 5 = -3 + 5 \qquad \text{Add 5 to each side.}$$

$$x = 2$$

Example:

$$-2x + -5 = 6$$

$$-2x + -5 - -5 = 6 - -5 \qquad \text{Subtract -5 from each side.}$$

$$-2x = 11$$

$$-2x \div -2 = 11 \div -2 \qquad \text{Divide each side by -2.}$$

$$x = -\frac{11}{2}$$

Doing the same thing to each side maintains equality even if it is not helpful to solving for the unknown amount. For example, we could take the equation $-3x + 7 = -8$ and add -2 to each side.

$$-3x + 7 = -8$$

$$-3x + 7 + -2 = -8 + -2 \qquad \text{Add -2 to each side.}$$

$$-3x + 5 = -10$$

If $-3x + 7 = -8$ is true then $-3x + 5 = -10$ is also true, but we are no closer to a solution than we were before adding -2. We can use moves that maintain equality to make new equations that all have the same solution. Helpful combinations of moves will eventually lead to an equation like $x = 5$, which gives the solution to the original equation (and every equation we wrote in the process of solving).

Dealing with Negative Numbers

1. Solve each equation.

   a. $4x = -28$

   b. $x - -6 = -2$

   c. $-x + 4 = -9$

   d. $-3x + 7 = 1$

   e. $25x + -11 = -86$

2. Here is an equation $2x + 9 = -15$. Write three different equations that have the same solution as $2x + 9 = -15$. Show or explain how you found them.

3. Select **all** the equations that match the diagram. (Lesson 6-3)

18

$x + 5$	$x + 5$	$x + 5$

   A. $x + 5 = 18$

   B. $18 \div 3 = x + 5$

   C. $3(x + 5) = 18$

   D. $x + 5 = \frac{1}{3} \cdot 18$

   E. $3x + 5 = 18$

NAME _____ DATE _____ PERIOD _____

4. There are 88 seats in a theater. The seating in the theater is split into 4 identical sections. Each section has 14 red seats and some blue seats.

(Lesson 6-2)

a. Draw a tape diagram to represent the situation.

b. What unknown amounts can be found by using the diagram or reasoning about the situation?

5. Match each story to an equation. (Lesson 6-4)

Stories

   a. A stack of nested paper cups is 8 inches tall. The first cup is 4 inches tall and each of the rest of the cups in the stack adds $\frac{1}{4}$ inch to the height of the stack.

   b. A baker uses 4 cups of flour. She uses $\frac{1}{4}$ cup to flour the counters and the rest to make 8 identical muffins.

   c. Elena has an 8-foot piece of ribbon. She cuts off a piece that is $\frac{1}{4}$ of a foot long and cuts the remainder into four pieces of equal length.

Equations

$$\frac{1}{4} + 4x = 8$$

$$4 + \frac{1}{4}x = 8$$

$$8x + \frac{1}{4} = 4$$

Lesson 6-10

Different Options for Solving One Equation

NAME _____ DATE _____ PERIOD _____

Learning Goal Let's think about which way is easier when we solve
equations with parentheses.

 ## Warm Up

10.1 Algebra Talk: Solve Each Equation

1. $100(x - 3) = 1,000$

2. $500(x - 3) = 5,000$

3. $0.03(x - 3) = 0.3$

4. $0.72(x + 2) = 7.2$

 ## Activity

10.2 Analyzing Solution Methods

Three students each attempted to solve the equation $2(x - 9) = 10$, but got
different solutions. Here are their methods. Do you agree with any of their
methods, and why?

Noah's method:

$2(x - 9) = 10$

$2(x - 9) + 9 = 10 + 9$ Add 9 to each side.

$2x = 19$

$2x \div 2 = 19 \div 2$ Divide each side by 2.

$x = \dfrac{19}{2}$

Elena's method:

$2(x - 9) = 10$

$2x - 18 = 10$ Apply the distributive property.

$2x - 18 - 18 = 10 - 18$ Subtract 18 from each side.

$2x = -8$

$2x \div 2 = \text{-}8 \div 2$ Divide each side by 2.

$x = \text{-}4$

Andre's method:

$2(x - 9) = 10$

$2x - 18 = 10$ Apply the distributive property.

$2x - 18 + 18 = 10 + 18$ Add 18 to each side.

$2x = 28$

$2x \div 2 = 28 \div 2$ Divide each side by 2.

$x = 14$

 ## Activity

10.3 Solution Pathways

For each equation, try to solve the equation using each method (dividing each side first, or applying the distributive property first). Some equations are easier to solve by one method than the other. When that is the case, stop doing the harder method and write down the reason you stopped.

1. $2{,}000(x - 0.03) = 6{,}000$

2. $2(x + 1.25) = 3.5$

3. $\frac{1}{4}(4 + x) = \frac{4}{3}$

4. $\text{-}10(x - 1.7) = \text{-}3$

5. $5.4 = 0.3(x + 8)$

NAME _____ DATE _____ PERIOD _____

Summary
Different Options for Solving One Equation

Equations can be solved in many ways. In this lesson, we focused on equations with a specific structure, and two specific ways to solve them.

Suppose we are trying to solve the equation $\frac{4}{5}(x + 27) = 16$. Two useful approaches are:

• divide each side by $\frac{4}{5}$

• apply the distributive property

In order to decide which approach is better, we can look at the numbers and think about which would be easier to compute. We notice that $\frac{4}{5} \cdot 27$ will be hard, because 27 isn't divisible by 5. But $16 \div \frac{4}{5}$ gives us $16 \cdot \frac{5}{4}$, and 16 is divisible by 4.

Dividing each side by $\frac{4}{5}$ gives:

$$\frac{4}{5}(x + 27) = 16$$

$$\frac{5}{4} \cdot \frac{4}{5}(x + 27) = 16 \cdot \frac{5}{4}$$

$$x + 27 = 20$$

$$x = -7$$

Sometimes the calculations are simpler if we first use the distributive property.

Let's look at the equation $100(x + 0.06) = 21$. If we first divide each side by 100, we get $\frac{21}{100}$ or 0.21 on the right side of the equation. But if we use the distributive property first, we get an equation that only contains whole numbers.

$$100(x + 0.06) = 21$$

$$100x + 6 = 21$$

$$100x = 15$$

$$x = \frac{15}{100}$$

Practice

Different Options for Solving One Equation

1. Andre wants to buy a backpack. The normal price of the backpack is $40. He notices that a store that sells the backpack is having a 30% off sale. What is the sale price of the backpack? **(Lesson 4-11)**

2. On the first math exam, 16 students received an A grade. On the second math exam, 12 students received an A grade. What percentage decrease is that? **(Lesson 4-12)**

3. Solve each equation.

   a. $2(x - 3) = 14$

   b. $-5(x - 1) = 40$

   c. $12(x + 10) = 24$

NAME _____ DATE _____ PERIOD _____

d. $\frac{1}{6}(x + 6) = 11$

e. $\frac{5}{7}(x - 9) = 25$

4. Select **all** expressions that represent a correct solution to the equation
   $6(x + 4) = 20$.

   (A.) $(20 - 4) \div 6$

   (B.) $\frac{1}{6}(20 - 4)$

   (C.) $20 - 6 - 4$

   (D.) $20 \div 6 - 4$

   (E.) $\frac{1}{6}(20 - 24)$

   (F.) $(20 - 24) \div 6$

5. Lin and Noah are solving the equation $7(x + 2) = 91$.

Lin starts by using the distributive property. Noah starts by dividing each side by 7.

  a. Show what Lin's and Noah's full solution methods might look like.

  b. What is the same and what is different about their methods?

Lesson 6-11

Using Equations to Solve Problems

NAME _____ DATE _____ PERIOD _____

Learning Goal Let's use tape diagrams, equations, and reasoning
to solve problems.

 ## Warm Up
11.1 Remember Tape Diagrams

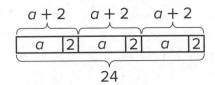

1. Write a story that could be represented by this tape diagram.

2. Write an equation that could be represented by this tape diagram.

1. Tyler is making invitations to the fair. He has already made some of the invitations, and he wants to finish the rest of them within a week. He is trying to spread out the remaining work, to make the same number of invitations each day. Tyler draws a diagram to represent the situation.

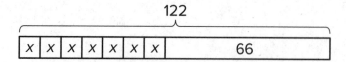

    a. Explain how each part of the situation is represented in Tyler's diagram.

- how many total invitations Tyler is trying to make

- how many invitations he has made already

- how many days he has to finish the invitations

    b. How many invitations should Tyler make each day to finish his goal within a week? Explain or show your reasoning.

    c. Use Tyler's diagram to write an equation that represents the situation. Explain how each part of the situation is represented in your equation.

    d. Show how to solve your equation.

NAME _____ DATE _____ PERIOD _____

2. Noah and his sister are making prize bags for a game at the fair. Noah is putting 7 pencil erasers in each bag. His sister is putting in some number of stickers. After filling 3 of the bags, they have used a total of 57 items.

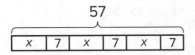

57

| x | 7 | x | 7 | x | 7 |

a. Explain how the diagram represents the situation.

b. Noah writes the equation $3(x + 7) = 57$ to represent the situation. Do you agree with him? Explain your reasoning.

c. How many stickers is Noah's sister putting in each prize bag? Explain or show your reasoning.

3. A family of 6 is going to the fair. They have a coupon for $1.50 off each ticket. If they pay $46.50 for all their tickets, how much does a ticket cost without the coupon? Explain or show your reasoning. If you get stuck, consider drawing a diagram or writing an equation.

Activity

11.3 Running Around

Priya, Han, and Elena, are members of the running club at school.

1. Priya was busy studying this week and ran 7 fewer miles than last week. She ran 9 times as far as Elena ran this week. Elena only had time to run 4 miles this week.

   a. How many miles did Priya run last week?

   b. Elena wrote the equation $\frac{1}{9}(x - 7) = 4$ to describe the situation. She solved the equation by multiplying each side by 9 and then adding 7 to each side. How does her solution compare to the way you found Priya's miles?

2. One day last week, 6 teachers joined $\frac{5}{7}$ of the members of the running club in an after-school run. Priya counted a total of 31 people running that day. How many members does the running club have?

3. Priya and Han plan a fundraiser for the running club. They begin with a balance of -80 because of expenses. In the first hour of the fundraiser they collect equal donations from 9 family members, which brings their balance to -44. How much did each parent give?

4. The running club uses the money they raised to pay for a trip to a canyon. At one point during a run in the canyon, the students are at an elevation of 128 feet. After descending at a rate of 50 feet per minute, they reach an elevation of -472 feet. How long did the descent take?

NAME _____ DATE _____ PERIOD _____

Are you ready for more?

A musician performed at three local fairs. At the first he doubled his money and spent $30. At the second he tripled his money and spent $54. At the third, he quadrupled his money and spent $72. In the end, he had $48 left. How much did he have before performing at the fairs?

Summary

Using Equations to Solve Problems

Many problems can be solved by writing and solving an equation.
Here is an example:

Clare ran 4 miles on Monday. Then for the next six days, she ran the same distance each day. She ran a total of 22 miles during the week. How many miles did she run on each of the 6 days?

One way to solve the problem is to represent the situation with an equation, $4 + 6x = 22$, where x represents the distance, in miles, she ran on each of the 6 days. Solving the equation gives the solution to this problem.

$$4 + 6x = 22$$
$$6x = 18$$
$$x = 3$$

Practice

Using Equations to Solve Problems

1. Find the value of each variable. (Lesson 5-9)

   a. $a \cdot 3 = -30$

   b. $-9 \cdot b = 45$

   c. $-89 \cdot 12 = c$

   d. $d \cdot 88 = -88{,}000$

2. Match each equation to its solution and to the story it describes.

   a. $5x - 7 = 3$

   b. $7 = 3(5 + x)$

   c. $3x + 5 = -7$

   d. $\frac{1}{3}(x + 7) = 5$

   Solutions

   -4

   $\frac{-8}{3}$

   2

   8

   Stories

   The temperature is -7. Since midnight the temperature tripled and then rose 5 degrees. What was the temperature at midnight?

   Jada has 7 pink roses and some white roses. She gives all of them away: 5 roses to each of her 3 favorite teachers. How many white roses did she give away?

   A musical instrument company reduced the time it takes for a worker to build a guitar. Before the reduction it took 5 hours. Now in 7 hours they can build 3 guitars. By how much did they reduce the time it takes to build each guitar?

   A club puts its members into 5 groups for an activity. After 7 students have to leave early, there are only 3 students left to finish the activity. How many students were in each group?

NAME _____ DATE _____ PERIOD _____

3. The baby giraffe weighed 132 pounds at birth. He gained weight at a steady rate for the first 7 months until his weight reached 538 pounds. How much did he gain each month?

4. Six teams are out on the field playing soccer. The teams all have the same number of players. The head coach asks for 2 players from each team to come help him move some equipment. Now there are 78 players on the field. Write and solve an equation whose solution is the number of players on each team.

5. A small town had a population of 960 people last year. The population grew to 1200 people this year. By what percentage did the population grow? **(Lesson 4-7)**

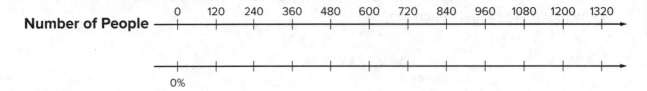

Number of People

| 0 | 120 | 240 | 360 | 480 | 600 | 720 | 840 | 960 | 1080 | 1200 | 1320 |

0%

6. The gas tank of a truck holds 30 gallons. The gas tank of a passenger car holds 50% less. How many gallons does it hold? **(Lesson 4-7)**

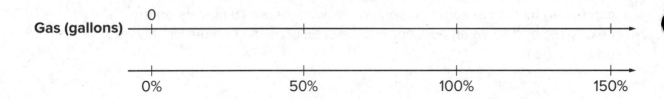

Gas (gallons)

0

0%   50%   100%   150%

Lesson 6-12

Solving Problems about Percent Increase or Decrease

NAME _____ DATE _____ PERIOD _____

Learning Goal Let's use tape diagrams, equations, and reasoning to solve problems with negatives and percents.

 Warm Up

12.1 20% Off

An item costs x dollars and then a 20% discount is applied. Select **all** the expressions that could represent the price of the item after the discount.

(A.) $\frac{20}{100}x$

(D.) $\frac{100 - 20}{100}x$

(B.) $x - \frac{20}{100}x$

(E.) $0.80x$

(C.) $(1 - 0.20)x$

(F.) $(100 - 20)x$

 Activity

12.2 Walking More Each Day

1. Mai started a new exercise program. On the second day, she walked 5 minutes more than on the first day. On the third day, she increased her walking time from day 2 by 20% and walked for 42 minutes. Mai drew a diagram to show her progress.

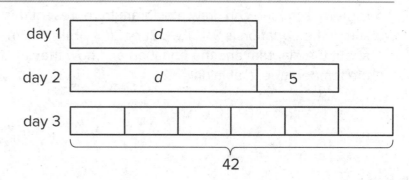

Explain how the diagram represents the situation.

2. Noah said the equation $1.20(d + 5) = 42$ also represents the situation. Do you agree with Noah? Explain your reasoning.

3. Find the number of minutes Mai walked on the first day. Did you use the diagram, the equation, or another strategy? Explain or show your reasoning.

4. Mai has been walking indoors because of cold temperatures. On Day 4 at noon, Mai hears a report that the temperature is only 9 degrees Fahrenheit. She remembers the morning news reporting that the temperature had doubled since midnight and was expected to rise 15 degrees by noon. Mai is pretty sure she can draw a diagram to represent this situation but isn't sure if the equation is $9 = 15 + 2t$ or $2(t + 15) = 9$. What would you tell Mai about the diagram and the equation and how they might be useful to find the temperature, t, at midnight?

NAME _____ DATE _____ PERIOD _____

Activity

12.3 A Sale on Shoes

1. A store is having a sale where all shoes are discounted by 20%. Diego has a coupon for $3 off of the regular price for one pair of shoes. The store first applies the coupon and then takes 20% off of the reduced price. If Diego pays $18.40 for a pair of shoes, what was their original price before the sale and without the coupon?

2. Before the sale, the store had 100 pairs of flip flops in stock. After selling some, they notice that $\frac{3}{5}$ of the flip flops they have left are blue. If the store has 39 pairs of blue flip flops, how many pairs of flip flops (any color) have they sold?

3. When the store had sold $\frac{2}{9}$ of the boots that were on display, they brought out another 34 pairs from the stock room. If that gave them 174 pairs of boots out, how many pairs were on display originally?

4. On the morning of the sale, the store donated 50 pairs of shoes to a homeless shelter. Then they sold 64% of their remaining inventory during the sale. If the store had 288 pairs after the donation and the sale, how many pairs of shoes did they have at the start?

A coffee shop offers a special: 33% extra free or 33% off the regular price. Which offer is a better deal? Explain your reasoning.

Summary

Solving Problems about Percent Increase or Decrease

We can solve problems where there is a percent increase or decrease by using what we know about equations.

For example, a camping store increases the price of a tent by 25%. A customer then uses a $10 coupon for the tent and pays $152.50.

We can draw a diagram that shows first the 25% increase and then the $10 coupon.

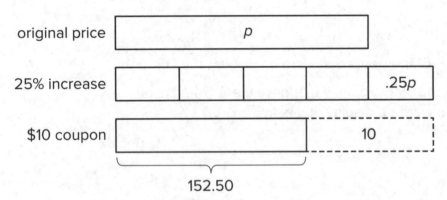

The price after the 25% increase is $p + .25p$ or $1.25p$. An equation that represents the situation could be $1.25p - 10 = 152.50$. To find the original price before the increase and discount, we can add 10 to each side and divide each side by 1.25, resulting in $p = 130$.

The original price of the tent was $130.

NAME _____ DATE _____ PERIOD _____

Practice
Solving Problems about Percent Increase or Decrease

1. A backpack normally costs $25 but it is on sale for $21. What percentage is the discount? **(Lesson 4-12)**

2. Find each product. **(Lesson 5-9)**

   a. $\frac{2}{5} \cdot (-10)$

   b. $-8 \cdot \left(\frac{-3}{2}\right)$

   c. $\frac{10}{6} \cdot 0.6$

   d. $\left(\frac{-100}{37}\right) \cdot (-0.37)$

3. Select **all** expressions that show x increased by 35%.

   (A.) $1.35x$

   (B.) $\frac{35}{100}x$

   (C.) $x + \frac{35}{100}x$

   (D.) $(1 + 0.35)x$

   (E.) $\frac{100 + 35}{100}x$

   (F.) $(100 + 35)x$

4. Complete each sentence with the word *discount*, *deposit*, or *withdrawal*.
(Lesson 4-11)

a. Clare took $20 out of her bank account. She made a _____.

b. Kiran used a coupon when he bought a pair of shoes.

He got a _____.

c. Priya put $20 into her bank account. She made a _____.

d. Lin paid less than usual for a pack of gum because it was on sale.

She got a _____.

5. Here are two stories.

• The initial freshman class at a college is 10% smaller than last year's class. But then during the first week of classes, 20 more students enroll. There are then 830 students in the freshman class.

• A store reduces the price of a computer by $20. Then during a 10% off sale, a customer pays $830.

Here are two equations.

• $0.9x + 20 = 830$

• $0.9(x - 20) = 830$

a. Decide which equation represents each story.

b. Explain why one equation has parentheses and the other doesn't.

c. Solve each equation, and explain what the solution means in the situation.

Lesson 6-13

Reintroducing Inequalities

NAME _____ DATE _____ PERIOD _____

Learning Goal Let's work with inequalities.

 ## Warm Up
13.1 Greater Than One

The number line shows values of x that make the inequality $x > 1$ true.

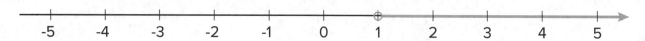

1. Select **all** the values of x from this list that make the inequality $x > 1$ true.

   (A.) 3

   (B.) -3

   (C.) 1

   (D.) 700

   (E.) 1.05

2. Name two more values of x that are solutions to the inequality.

A sign next to a roller coaster at an amusement park says, "You must be at least 60 inches tall to ride." Noah is happy to know that he is tall enough to ride.

1. Noah is x inches tall. Which of the following can be true: $x > 60$, $x = 60$, or $x < 60$? Explain how you know.

2. Noah's friend is 2 inches shorter than Noah. Can you tell if Noah's friend is tall enough to go on the ride? Explain or show your reasoning.

3. List one possible height for Noah that means that his friend is tall enough to go on the ride, and another that means that his friend is too short for the ride.

Marcio Jose Bastos Silva/Shutterstock

NAME _____ DATE _____ PERIOD _____

4. On the number line below, show all the possible heights that Noah's friend could be.

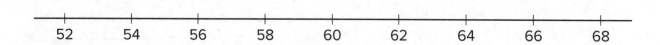

5. Noah's friend is y inches tall. Use y and any of the symbols $<$, $=$, $>$ to express this height.

 ## Activity

13.3 Is the Inequality True or False?

The table shows four inequalities and four possible values for x. Decide whether each value makes each inequality true, and complete the table with "true" or "false." Discuss your thinking with your partner. If you disagree, work to reach an agreement.

x	0	100	-100	25
$x \leq 25$				
$100 < 4x$				
$-3x > -75$				
$10 \geq 35 - x$				

Find an example of an inequality used in the real world and describe it using a number line.

Summary
Reintroducing Inequalities

We use inequalities to describe a range of numbers.

In many places, you are allowed to get a driver's license when you are at least 16 years old. When checking if someone is old enough to get a license, we want to know if their age is at least 16.

If h is the age of a person, then we can check if they are allowed to get a driver's license by checking if their age makes the inequality $h > 16$ (they are older than 16) or the equation $h = 16$ (they are 16) true.

The symbol $\geq$, pronounced "greater than or equal to," combines these two cases and we can just check if $h \geq 16$ (their age is greater than or equal to 16). The inequality $h \geq 16$ can be represented on a number line:

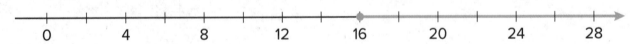

NAME _____ DATE _____ PERIOD _____

 Practice

Reintroducing Inequalities

1. For each inequality, find two values for x that make the inequality true and two values that make it false.

   a. $x + 3 > 70$

   b. $x + 3 < 70$

   c. $-5x < 2$

   d. $5x < 2$

2. Here is an inequality: $-3x > 18$.

   a. List some values for x that would make this inequality true.

   b. How are the solutions to the inequality $-3x \geq 18$ different from the solutions to $-3x > 18$? Explain your reasoning.

3. Here are the prices for cheese pizza at a certain pizzeria: (Lesson 4-12)

Pizza Size	Price in Dollars
Small	11.60
Medium	
Large	16.25

  a. You had a coupon that made the price of a large pizza $13.00. For what percent off was the coupon?

  b. Your friend purchased a medium pizza for $10.31 with a 30% off coupon. What is the price of a medium pizza without a coupon?

  c. Your friend has a 15% off coupon and $10. What is the largest pizza that your friend can afford, and how much money will be left over after the purchase?

4. Select **all** the stories that can be represented by the diagram. (Lesson 6-4)

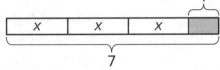

  (A.) Andre studies 7 hours this week for end-of-year exams. He spends 1 hour on English and an equal number of hours each on math, science, and history.

  (B.) Lin spends $3 on 7 markers and a $1 pen.

  (C.) Diego spends $1 on 7 stickers and 3 marbles.

  (D.) Noah shares 7 grapes with 3 friends. He eats 1 and gives each friend the same number of grapes.

  (E.) Elena spends $7 on 3 notebooks and a $1 pen.

Lesson 6-14

Finding Solutions to Inequalities in Context

NAME _____ DATE _____ PERIOD _____

Learning Goal Let's solve more complicated inequalities.

 ## Warm Up
14.1 Solutions to Equations and Solutions to Inequalities

1. Solve $-x = 10$.

2. Find 2 solutions to $-x > 10$.

3. Solve $2x = -20$.

4. Find 2 solutions to $2x > -20$.

 ## Activity
14.2 Earning Money for Soccer Stuff

1. Andre has a summer job selling magazine subscriptions. He earns $25 per week plus $3 for every subscription he sells. Andre hopes to make at least enough money this week to buy a new pair of soccer cleats.

   a. Let n represent the number of magazine subscriptions Andre sells this week. Write an expression for the amount of money he makes this week.

   b. The least expensive pair of cleats Andre wants costs $68. Write and solve an equation to find out how many magazine subscriptions Andre needs to sell to buy the cleats.

c. If Andre sold 16 magazine subscriptions this week, would he reach his goal? Explain your reasoning.

d. What are some other numbers of magazine subscriptions Andre could have sold and still reached his goal?

e. Write an *inequality* expressing that Andre wants to make at least $68.

f. Write an inequality to describe the number of subscriptions Andre must sell to reach his goal.

2. Diego has budgeted $35 from his summer job earnings to buy shorts and socks for soccer. He needs 5 pairs of socks and a pair of shoes. The socks cost different amounts in different stores. The shorts he wants cost $19.95.

a. Let x represent the price of one pair of socks. Write an expression for the total cost of the socks and shorts.

b. Write and solve an equation that says that Diego spent exactly $35 on the socks and shorts.

NAME _____ DATE _____ PERIOD _____

c. List some other possible prices for the socks that would still allow Diego to stay within his budget.

d. Write an inequality to represent the amount Diego can spend on a single pair of socks.

 Activity

14.3 Granola Bars and Savings

1. Kiran has $100 saved in a bank account. (The account doesn't earn interest.) He asked Clare to help him figure out how much he could take out each month if he needs to have at least $25 in the account a year from now.

a. Clare wrote the inequality -12x + 100 ≥ 25, where x represents the amount Kiran takes out each month. What does -12x represent?

b. Find some values of x that would work for Kiran.

c. We could express *all* the values that would work using either x ≤ ____ or x ≥ ____. Which one should we use?

d. Write the answer to Kiran's question using mathematical notation.

2. A teacher wants to buy 9 boxes of granola bars for a school trip. Each box usually costs $7, but many grocery stores are having a sale on granola bars this week. Different stores are selling boxes of granola bars at different discounts.

   a. If x represents the dollar amount of the discount, then the amount the teacher will pay can be expressed as $9(7 - x)$. In this expression, what does the quantity $7 - x$ represent?

   b. The teacher has $36 to spend on the granola bars. The equation $9(7 - x) = 36$ represents a situation where she spends all $36. Solve this equation.

   c. What does the solution mean in this situation?

   d. The teacher does not have to spend all $36. Write an inequality relating 36 and $9(7 - x)$ representing this situation.

   e. The solution to this inequality must either look like $x \geq 3$ or $x \leq 3$. Which do you think it is? Explain your reasoning.

NAME _____ DATE _____ PERIOD _____

Are you ready for more?

Jada and Diego baked a large batch of cookies.

- They selected $\frac{1}{4}$ of the cookies to give to their teachers.

- Next, they threw away one burnt cookie.

- They delivered $\frac{2}{5}$ of the remaining cookies to a local nursing home.

- Next, they gave 3 cookies to some neighborhood kids.

- They wrapped up $\frac{2}{3}$ of the remaining cookies to save for their friends.

After all this, they had 15 cookies left. How many cookies did they bake?

Summary

Finding Solutions to Inequalities in Context

Suppose Elena has $5 and sells pens for $1.50 each. Her goal is to save $20.

We could solve the equation $1.5x + 5 = 20$ to find the number of pens, x, that Elena needs to sell in order to save *exactly* $20. Adding -5 to both sides of the equation gives us $1.5x = 15$, and then dividing both sides by 1.5 gives the solution $x = 10$ pens.

What if Elena wants to have some money left over?

The inequality $1.5x + 5 > 20$ tells us that the amount of money Elena makes needs to be *greater* than $20.

The solution to the previous equation will help us understand what the solutions to the inequality will be. We know that if she sells 10 pens, she will make $20. Since each pen gives her more money, she needs to sell *more* than 10 pens to make more than $20. So the **solution to the inequality** is $x > 10$.

Glossary

solution to an inequality

1. The solution to $5 - 3x > 35$ is either $x > -10$ or $-10 > x$. Which solution is correct? Explain how you know.

2. The school band director determined from past experience that if they charge t dollars for a ticket to the concert, they can expect attendance of $1,000 - 50t$. The director used this model to figure out that the ticket price needs to be $8 or greater in order for at least 600 to attend. Do you agree with this claim? Why or why not?

3. Which inequality is true when the value of x is -3? **(Lesson 6-13)**

   (A.) $-x - 6 < -3.5$

   (B.) $-x - 6 > 3.5$

   (C.) $-x - 6 > -3.5$

   (D.) $x - 6 > -3.5$

NAME _____ DATE _____ PERIOD _____

4. Draw the solution set for each of the following inequalities. **(Lesson 6-13)**

a. $x \leq 5$

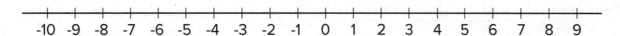

b. $x < \dfrac{5}{2}$

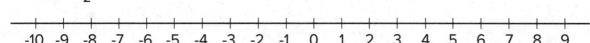

5. Write three different equations that match the tape diagram. **(Lesson 6-3)**

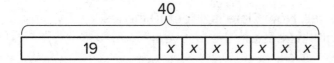

6. A baker wants to reduce the amount of sugar in his cake recipes. He decides to reduce the amount used in 1 cake by $\frac{1}{2}$ cup. He then uses $4\frac{1}{2}$ cups of sugar to bake 6 cakes. (Lesson 6-2)

$$\underbrace{\boxed{x - \tfrac{1}{2}}\boxed{x - \tfrac{1}{2}}\boxed{x - \tfrac{1}{2}}\boxed{x - \tfrac{1}{2}}\boxed{x - \tfrac{1}{2}}\boxed{x - \tfrac{1}{2}}}_{4\frac{1}{2}}$$

a. Describe how the tape diagram represents the story.

b. How much sugar was originally in each cake recipe?

7. One year ago, Clare was 4 feet 6 inches tall. Now Clare is 4 feet 10 inches tall. By what percentage did Clare's height increase in the last year?
(Lesson 4-12)

Lesson 6-15

Efficiently Solving Inequalities

NAME _____ DATE _____ PERIOD _____

Learning Goal Let's solve more complicated inequalities.

 ## Warm Up
15.1 Lots of Negatives

Here is an inequality: $-x \geq -4$.

1. Predict what you think the solutions on the number line will look like.

2. Select **all** the values that are solutions to $-x \geq -4$.

    (A.) 3   (D.) -4

    (B.) -3   (E.) 4.001

    (C.) 4   (F.) -4.001

3. Graph the solutions to the inequality on the number line.

    ┼──┼──┼──┼──┼──┼──┼──┼──┼──┼──┼──┼──┼──┼

Activity

15.2 Inequalities with Tables

1. Let's investigate the inequality $x - 3 > -2$.

x	-4	-3	-2	-1	0	1	2	3	4
x − 3	-7		-5				-1		1

   a. Complete the table.

   b. For which values of x is it true that $x - 3 = -2$?

   c. For which values of x is it true that $x - 3 > -2$?

   d. Graph the solutions to $x - 3 > -2$ on the number line.

2. Here is an inequality: $2x < 6$.

   a. Predict which values of x will make the inequality $2x < 6$ true.

   b. Complete the table. Does it match your prediction?

x	-4	-3	-2	-1	0	1	2	3	4
2x									

   c. Graph the solutions to $2x < 6$ on the number line.

NAME _____ DATE _____ PERIOD _____

3. Here is an inequality: $-2x < 6$.

    a. Predict which values of x will make the inequality $-2x < 6$ true.

    b. Complete the table. Does it match your prediction?

x	-4	-3	-2	-1	0	1	2	3	4
-2x									

    c. Graph the solutions to $-2x < 6$ on the number line.

    d. How are the solutions to $2x < 6$ different from the solutions to $-2x < 6$?

Activity

15.3 Which Side are the Solutions?

1. Let's investigate $-4x + 5 \geq 25$.

    a. Solve $-4x + 5 = 25$.

    b. Is $-4x + 5 \geq 25$ true when x is 0? What about when x is 7? What about when x is -7?

    c. Graph the solutions to $-4x + 5 \geq 25$ on the number line.

2. Let's investigate $\frac{4}{3}x + 3 < \frac{23}{3}$.

    a. Solve $\frac{4}{3}x + 3 = \frac{23}{3}$.

    b. Is $\frac{4}{3}x + 3 < \frac{23}{3}$ true when x is 0?

    c. Graph the solutions to $\frac{4}{3}x + 3 < \frac{23}{3}$ on the number line.

3. Solve the inequality $3(x + 4) > 17.4$ and graph the solutions on the number line.

4. Solve the inequality $-3\left(x - \frac{4}{3}\right) \leq 6$ and graph the solutions on the number line.

NAME _____ DATE _____ PERIOD _____

Are you ready for more?

Write at least three different inequalities whose solution is $x > \text{-}10$.
Find one with x on the left side that uses a $<$.

Summary

Efficiently Solving Inequalities

Here is an inequality: $3(10 - 2x) < 18$. The solution to this inequality is all the values you could use in place of x to make the inequality true.

In order to solve this, we can first solve the related equation $3(10 - 2x) = 18$ to get the solution $x = 2$. That means 2 is the boundary between values of x that make the inequality true and values that make the inequality false.

To solve the inequality, we can check numbers greater than 2 and less than 2 and see which ones make the inequality true.

Let's check a number that is greater than 2: $x = 5$.

- Replacing x with 5 in the inequality, we get $3(10 - 2 \cdot 5) < 18$ or just $0 < 18$.

- This is true, so $x = 5$ is a solution.

- This means that all values greater than 2 make the inequality true. We can write the solutions as $x > 2$ and also represent the solutions on a number line:

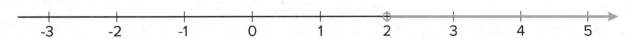

Notice that 2 itself is not a solution because it's the value of x that makes $3(10 - 2x)$ *equal* to 18, and so it does not make $3(10 - 2x) < 18$ true.

For confirmation that we found the correct solution, we can also test a value that is less than 2. If we test $x = 0$, we get $3(10 - 2 \cdot 0) < 18$ or just $30 < 18$. This is false, so $x = 0$ and all values of x that are less than 2 are not solutions.

Practice

Efficiently Solving Inequalities

1. Respond to each of the following.

   a. Consider the inequality $-1 \leq \frac{x}{2}$.

      i. Predict which values of x will make the inequality true.

      ii. Complete the table to check your prediction.

x	-4	-3	-2	-1	0	1	2	3	4
$\frac{x}{2}$									

   b. Consider the inequality $1 \leq \frac{-x}{2}$.

      i. Predict which values of x will make it true.

      ii. Complete the table to check your prediction.

x	-4	-3	-2	-1	0	1	2	3	4
$\frac{-x}{2}$	2		1		0		-1		-2

2. Diego is solving the inequality $100 - 3x \geq -50$. He solves the equation $100 - 3x = -50$ and gets $x = 50$. What is the solution to the inequality?

   (A.) $x < 50$

   (B.) $x \leq 50$

   (C.) $x > 50$

   (D.) $x \geq 50$

NAME _____ DATE _____ PERIOD _____

3. Solve the inequality -5(x − 1) > -40, and graph the solution on a number line.

4. Select **all** values of x that make the inequality -x + 6 ≥ 10 true. (Lesson 6-13)

   Ⓐ -3.9   Ⓔ 4.01

   Ⓑ 4   Ⓕ 3.9

   Ⓒ -4.01   Ⓖ 0

   Ⓓ -4   Ⓗ -7

5. Draw the solution set for each of the following inequalities. (Lesson 6-13)

   a. x > 7

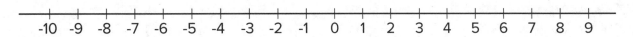

   b. x ≥ -4.2

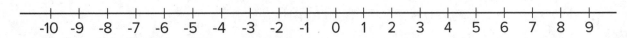

6. The price of a pair of earrings is $22 but Priya buys them on sale for $13.20. (Lesson 4-12)

   a. By how much was the price discounted?

   b. What was the percentage of the discount?

Lesson 6-16

Interpreting Inequalities

NAME _____ DATE _____ PERIOD _____

Learning Goal Let's write inequalities.

 ## Warm Up

16.1 Solve Some Inequalities!

For each inequality, find the value or values of x that make it true.

1. $8x + 21 \le 56$

2. $56 < 7(7 - x)$

 ## Activity

16.2 Club Activities Matching

Choose the inequality that best matches each given situation. Explain your reasoning.

1. The Garden Club is planting fruit trees in their school's garden. There is one large tree that needs 5 pounds of fertilizer. The rest are newly planted trees that need $\frac{1}{2}$ pound fertilizer each.

(A.) $25x + 5 \le \frac{1}{2}$

(B.) $\frac{1}{2}x + 5 \le 25$

(C.) $\frac{1}{2}x + 25 \le 5$

(D.) $5x + \frac{1}{2} \le 25$

2. The Chemistry Club is experimenting with different mixtures of water with a certain chemical (sodium polyacrylate) to make fake snow.

   To make each mixture, the students start with some amount of water, and then add $\frac{1}{7}$ of that amount of the chemical, and then 9 more grams of the chemical. The chemical is expensive, so there can't be more than a certain number of grams of the chemical in any one mixture.

   (A.) $\frac{1}{7}x + 9 \leq 26.25$

   (B.) $9x + \frac{1}{7} \leq 26.25$

   (C.) $26.25x + 9 \leq \frac{1}{7}$

   (D.) $\frac{1}{7}x + 26.25 \leq 9$

3. The Hiking Club is on a hike down a cliff. They begin at an elevation of 12 feet and descend at the rate of 3 feet per minute.

   (A.) $37x - 3 \geq 12$

   (B.) $3x - 37 \geq 12$

   (C.) $12 - 3x \geq -37$

   (D.) $12x - 37 \geq -3$

NAME _____ DATE _____ PERIOD _____

4. The Science Club is researching boiling points. They learn that at high altitudes, water boils at lower temperatures. At sea level, water boils at 212°F. With each increase of 500 feet in elevation, the boiling point of water is lowered by about 1°F.

(A.) $212 - \dfrac{1}{500}e < 195$

(B.) $\dfrac{1}{500}e - 195 < 212$

(C.) $195 - 212e < \dfrac{1}{500}$

(D.) $212 - 195e < \dfrac{1}{500}$

Activity

16.3 Club Activities Display

Your teacher will assign your group *one* of the situations from the last task. Create a visual display about your situation. In your display:

- Explain what the variable and each part of the inequality represent.

- Write a question that can be answered by the solution to the inequality.

- Show how you solved the inequality.

- Explain what the solution means in terms of the situation.

{3, 4, 5, 6} is a set of four consecutive integers whose sum is 18.

1. How many sets of three consecutive integers are there whose sum is between 51 and 60? Can you be sure you've found them all? Explain or show your reasoning.

2. How many sets of four consecutive integers are there whose sum is between 59 and 82? Can you be sure you've found them all? Explain or show your reasoning.

NAME _____ DATE _____ PERIOD _____

Summary
Interpreting Inequalities

We can represent and solve many real-world problems with inequalities. Writing the inequalities is very similar to writing equations to represent a situation. The expressions that make up the inequalities are the same as the ones we have seen in earlier lessons for equations.

For inequalities, we also have to think about how expressions compare to each other, which one is bigger, and which one is smaller. Can they also be equal?

For example, a school fundraiser has a minimum target of $500. Faculty have donated $100 and there are 12 student clubs that are participating with different activities. How much money should each club raise to meet the fundraising goal?

If n is the amount of money that each club raises, then the solution to $100 + 12n = 500$ is the minimum amount each club has to raise to meet the goal.

It is more realistic, though, to use the inequality $100 + 12n \geq 500$ since the more money we raise, the more successful the fundraiser will be. There are many solutions because there are many different amounts of money the clubs could raise that would get us above our minimum goal of $500.

1. Priya looks at the inequality $12 - x > 5$ and says "I subtract a number from 12 and want a result that is bigger than 5. That means that the solutions should be values of x that are smaller than something."

   Do you agree with Priya? Explain your reasoning and include solutions to the inequality in your explanation.

2. When a store had sold $\frac{2}{5}$ of the shirts that were on display, they brought out another 30 from the stockroom. The store likes to keep at least 150 shirts on display. The manager wrote the inequality $\frac{3}{5}x + 30 \geq 150$ to describe the situation.

   a. Explain what $\frac{3}{5}$ means in the inequality.

   b. Solve the inequality.

   c. Explain what the solution means in the situation.

NAME _____ DATE _____ PERIOD _____

3. You know *x* is a number less than 4. Select **all** the inequalities that *must* be true. (Lesson 6-13)

(A.) $x < 2$

(B.) $x + 6 < 10$

(C.) $5x < 20$

(D.) $x - 2 > 2$

(E.) $x < 8$

4. Here is an unbalanced hanger. (Lesson 6-13)

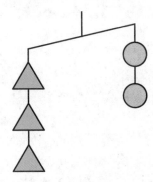

a. If you knew each circle weighed 6 grams, what would that tell you about the weight of each triangle? Explain your reasoning.

b. If you knew each triangle weighed 3 grams, what would that tell you about the weight of each circle? Explain your reasoning.

5. Match each sentence with the inequality that could represent the situation. (Lesson 6-13)

Situations	Inequalities
a. Han got $2 from Clare, but still has less than $20.	$x - 2 < 20$
	$2x < 20$
b. Mai spent $2 and has less than $20.	$x + 2 < 20$
	$\frac{1}{2}x < 20$
c. If Tyler had twice the amount of money he has, he would have less than $20.	
d. If Priya had half the money she has, she would have less than $20.	

6. At a skateboard shop: (Lesson 4-12)

a. The price tag on a shirt says $12.58. Sales tax is 7.5% of the price. How much will you pay for the shirt?

b. The store buys a helmet for $19.00 and sells it for $31.50. What percentage was the markup?

c. The shop pays workers $14.25 per hour plus 5.5% commission. If someone works 18 hours and sells $250 worth of merchandise, what is the total amount of their paycheck for this pay period? Explain or show your reasoning.

Lesson 6-17

Modeling with Inequalities

NAME _____ DATE _____ PERIOD _____

Learning Goal Let's look at solutions to inequalities.

Warm Up
17.1 Possible Values

The stage manager of the school musical is trying to figure out how many sandwiches he can order with the $83 he collected from the cast and crew. Sandwiches cost $5.99 each, so he lets x represent the number of sandwiches he will order and writes $5.99x \le 83$. He solves this to 2 decimal places, getting $x \le 13.86$.

Which of these are valid statements about this situation? (Select **all** that apply.)

(A.) He can call the sandwich shop and order exactly 13.86 sandwiches.

(B.) He can round up and order 14 sandwiches.

(C.) He can order 12 sandwiches.

(D.) He can order 9.5 sandwiches.

(E.) He can order 2 sandwiches.

(F.) He can order -4 sandwiches.

Activity

17.2 Elevator

A mover is loading an elevator with many identical 48-pound boxes.
The mover weighs 185 pounds. The elevator can carry at most 2,000 pounds.

1. Write an inequality that says that the mover will not overload the elevator on a particular ride. Check your inequality with your partner.

2. Solve your inequality and explain what the solution means.

3. Graph the solution to your inequality on a number line.

4. If the mover asked, "How many boxes can I load on this elevator at a time?" what would you tell them?

Activity

17.3 Info Gap: Giving Advice

Your teacher will give you either a *problem card* or a *data card*. Do not show or read your card to your partner.

If your teacher gives you the *problem card:*	If your teacher gives you the *data card:*
1. Silently read your card and think about what information you need to be able to answer the question.	1. Silently read your card.
2. Ask your partner for the specific information that you need.	2. Ask your partner *"What specific information do you need?"* and wait for them to *ask* for information.
3. Explain how you are using the information to solve the problem.	If your partner asks for information that is not on the card, do not do the calculations for them. Tell them you don't have that information.
Continue to ask questions until you have enough information to solve the problem.	3. Before sharing the information, ask *"Why do you need that information?"* Listen to your partner's reasoning and ask clarifying questions.
4. Share the *problem card* and solve the problem independently.	4. Read the *problem card* and solve the problem independently.
5. Read the *data card* and discuss your reasoning.	5. Share the *data card* and discuss your reasoning.

Pause here so your teacher can review your work. Ask your teacher for a new set of cards and repeat the activity, trading roles with your partner.

NAME _____ DATE _____ PERIOD _____

Are you ready for more?

In a day care group, nine babies are five months old and 12 babies are seven months old. How many full months from now will the average age of the 21 babies first surpass 20 months old?

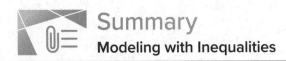

Summary
Modeling with Inequalities

We can represent and solve many real-world problems with inequalities.

Whenever we write an inequality, it is important to decide what quantity we are representing with a variable. After we make that decision, we can connect the quantities in the situation to write an expression, and finally, the whole inequality.

As we are solving the inequality or equation to answer a question, it is important to keep the meaning of each quantity in mind. This helps us to decide if the final answer makes sense in the context of the situation.

For example: Han has 50 centimeters of wire and wants to make a square picture frame with a loop to hang it that uses 3 centimeters for the loop.

This situation can be represented by $3 + 4s = 50$, where s is the length of each side (if we want to use all the wire).

We can also use $3 + 4s \leq 50$ if we want to allow for solutions that don't use all the wire. In this case, any positive number that is less or equal to 11.75 cm is a solution to the inequality. Each solution represents a possible side length for the picture frame since Han can bend the wire at any point.

In other situations, the variable may represent a quantity that increases by whole numbers, such as with numbers of magazines, loads of laundry, or students. In those cases, only whole-number solutions make sense.

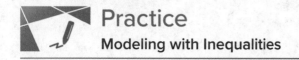

Practice

Modeling with Inequalities

1. 28 students travel on a field trip. They bring a van that can seat 12 students. Elena and Kiran's teacher asks other adults to drive cars that seat 3 children each to transport the rest of the students.

    Elena wonders if she should use the inequality $12 + 3n > 28$ or $12 + 3n \geq 28$ to figure out how many cars are needed. Kiran doesn't think it matters in this case. Do you agree with Kiran? Explain your reasoning.

2. Respond to each question.

    a. In the cafeteria, there is one large 10-seat table and many smaller 4-seat tables. There are enough tables to fit 200 students. Write an inequality whose solution is the possible number of 4-seat tables in the cafeteria.

    b. 5 barrels catch rainwater in the schoolyard. Four barrels are the same size, and the fifth barrel holds 10 liters of water. Combined, the 5 barrels can hold at least 200 liters of water. Write an inequality whose solution is the possible size of each of the 4 barrels.

    c. How are these two problems similar? How are they different?

NAME _____ DATE _____ PERIOD _____

3. Solve each equation. (Lesson 6-9)

a. $5(n - 4) = -60$

b. $-3t + -8 = 25$

c. $7p - 8 = -22$

d. $\frac{2}{5}(j + 40) = -4$

e. $4(w + 1) = -6$

4. Select **all** the inequalities that have the same graph as $x < 4$. (Lesson 6-13)

Ⓐ $x < 2$

Ⓑ $x + 6 < 10$

Ⓒ $5x < 20$

Ⓓ $x - 2 > 2$

Ⓔ $x < 8$

5. A 200 pound person weighs 33 pounds on the Moon. (Lesson 4-12)

   a. How much did the person's weight decrease?

   b. By what percentage did the person's weight decrease?

Lesson 6-18

Subtraction in Equivalent Expressions

NAME _____ DATE _____ PERIOD _____

Learning Goal Let's find ways to work with subtraction in expressions.

 ## Warm Up
18.1 Number Talk: Additive Inverses

Find each sum or difference mentally.

1. -30 + -10   **2.** -10 + -30   **3.** -30 − 10   **4.** 10 − -30

 ## Activity
18.2 A Helpful Observation

Lin and Kiran are trying to calculate $7\frac{3}{4} + 3\frac{5}{6} - 1\frac{3}{4}$. Here is their conversation.

Lin: "I plan to first add $7\frac{3}{4}$ and $3\frac{5}{6}$, so I will have to start by finding equivalent fractions with a common denominator."

Kiran: "It would be a lot easier if we could start by working with the $1\frac{3}{4}$ and $7\frac{3}{4}$. Can we rewrite it like $7\frac{3}{4} + 1\frac{3}{4} - 3\frac{5}{6}$?"

Lin: "You can't switch the order of numbers in a subtraction problem like you can with addition; 2 − 3 is not equal to 3 − 2."

Kiran: "That's true, but do you remember what we learned about rewriting subtraction expressions using addition? 2 − 3 is equal to 2 + (-3)."

1. Write an expression that is equivalent to $7\frac{3}{4} + 3\frac{5}{6} - 1\frac{3}{4}$ that uses addition instead of subtraction.

2. If you wrote the terms of your new expression in a different order, would it still be equivalent? Explain your reasoning.

Activity
18.3 Organizing Work

1. Write two expressions for the area of the big rectangle.

$\frac{1}{2}$
8y	x	12

2. Use the distributive property to write an expression that is equivalent to $\frac{1}{2}(8y + \text{-}x + \text{-}12)$. The boxes can help you organize your work.

$\frac{1}{2}$
8y	-x	-12

3. Use the distributive property to write an expression that is equivalent to $\frac{1}{2}(8y - x - 12)$.

NAME _____ DATE _____ PERIOD _____

Are you ready for more?

Here is a calendar for April 2017.

APRIL 2017						
SUNDAY	MONDAY	TUESDAY	WEDNESDAY	THURSDAY	FRIDAY	SATURDAY
						1
2	3	4	5	6	7	8
9	10	11	12	13	14	15
16	17	18	19	20	21	22
23	24	25	26	27	28	29
30						

Let's choose a date: the 10th. Look at the numbers above, below, and to either side of the 10th: 3, 17, 9, 11.

1. Average these four numbers. What do you notice?

2. Choose a different date that is in a location where it has a date above, below, and to either side. Average these four numbers. What do you notice?

3. Explain why the same thing will happen for any date in a location where it has a date above, below, and to either side.

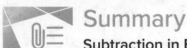

Summary
Subtraction in Equivalent Expressions

Working with subtraction and signed numbers can sometimes get tricky. We can apply what we know about the relationship between addition and subtraction—that subtracting a number gives the same result as adding its opposite—to our work with expressions.

Then, we can make use of the properties of addition that allow us to add and group in any order. This can make calculations simpler.

For example:

$$\frac{5}{8} - \frac{2}{3} - \frac{1}{8}$$

$$\frac{5}{8} + -\frac{2}{3} + -\frac{1}{8}$$

$$\frac{5}{8} + -\frac{1}{8} + -\frac{2}{3}$$

$$\frac{4}{8} + -\frac{2}{3}$$

We can also organize the work of multiplying signed numbers in expressions.

The product $\frac{3}{2}(6y - 2x - 8)$ can be found by drawing a rectangle with the first factor, $\frac{3}{2}$, on one side, and the three terms inside the parentheses on the other side.

	6y	-2x	-8
$\frac{3}{2}$			

Multiply $\frac{3}{2}$ by each term across the top and perform the multiplications.

	6y	-2x	-8
$\frac{3}{2}$	$\frac{3}{2} \cdot 6y$	$\frac{3}{2} \cdot -2x$	$\frac{3}{2} \cdot -8$

Reassemble the parts to get the expanded version of the original expression:

	6y	-2x	-8
$\frac{3}{2}$	9y	-3x	-12

$$\frac{3}{2}(6y - 2x - 8) = 9y - 3x - 12$$

Glossary

term

NAME _____ DATE _____ PERIOD _____

 ## Practice
Subtraction in Equivalent Expressions

1. For each expression, write an equivalent expression that uses only addition.

   a. $20 - 9 + 8 - 7$

   b. $4x - 7y - 5z + 6$

   c. $-3x - 8y - 4 - \frac{8}{7}z$

2. Use the distributive property to write an expression that is equivalent to each expression. If you get stuck, consider drawing boxes to help organize your work.

   a. $9\left(4x - 3y - \frac{2}{3}\right)$

   b. $-2(-6x + 3y - 1)$

   c. $\frac{1}{5}(20y - 4x - 13)$

   d. $8\left(-x - \frac{1}{2}\right)$

   e. $-8\left(-x - \frac{3}{4}y + \frac{7}{2}\right)$

3. Kiran wrote the expression $x - 10$ for this number puzzle: "Pick a number, add -2, and multiply by 5."

Lin thinks Kiran made a mistake.

   a. How can she convince Kiran he made a mistake?

   b. What would be a correct expression for this number puzzle?

4. The output from a coal power plant is shown in the table. (Lesson 2-7)

Energy in Megawatts	Number of Days
1,200	2.4
1,800	3.6
4,000	8
10,000	20

Similarly, the output from a solar power plant is shown in the table.

Energy in Megawatts	Number of Days
100	1
650	4
1,200	7
1,750	10

Based on the tables, is the energy output in proportion to the number of days for either plant? If so, write an equation showing the relationship. If not, explain your reasoning.

Lesson 6-19

Expanding and Factoring

NAME _____ DATE _____ PERIOD _____

Learning Goal Let's use the distributive property to write expressions in different ways.

 ## Warm Up
19.1 Number Talk: Parentheses

Find the value of each expression mentally.

1. $2 + 3 \cdot 4$

2. $(2 + 3)(4)$

3. $2 - 3 \cdot 4$

4. $2 - (3 + 4)$

Activity

19.2 Factoring and Expanding with Negative Numbers

In each row, write the equivalent expression. If you get stuck, use a diagram to organize your work. The first row in the left column is provided as an example. Diagrams are provided for the first three rows in the left column.

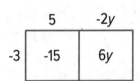

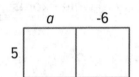

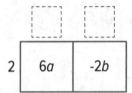

Factored	Expanded	Factored	Expanded
$-3(5 - 2y)$	$-15 + 6y$	$k(4 - 17)$	
$5(a - 6)$			$10a - 13a$
	$6a - 2b$	$-2x(3y - z)$	
$-4(2w - 5z)$			$ab - bc - 3bd$
$-(2x - 3y)$		$-x(3y - z + 4w)$	
	$20x - 10y + 15z$		

Are you ready for more?

Expand to create an equivalent expression that uses the fewest number of terms:

$\left(\left(\left((x + 1)\frac{1}{2}\right) + 1\right)\frac{1}{2}\right) + 1$. If we wrote a new expression following the same pattern so that there were 20 sets of parentheses, how could it be expanded into an equivalent expression that uses the fewest number of terms?

NAME _____ DATE _____ PERIOD _____

Summary
Expanding and Factoring

We can use properties of operations in different ways to rewrite expressions and create equivalent expressions. We have already seen that we can use the distributive property to expand an expression, for example $3(x + 5) = 3x + 15$. We can also use the distributive property in the other direction and factor an expression, for example $8x + 12 = 4(2x + 3)$.

We can organize the work of using the distributive property to rewrite the expression $12x - 8$. In this case we know the product and need to find the factors.

The terms of the product go inside.

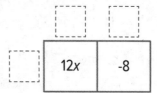

We look at the expressions and think about a factor they have in common. $12x$ and -8 each have a factor of 4. We place the common factor on one side of the large rectangle.

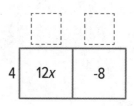

Now we think: "4 times *what* is $12x$?" "4 times *what* is -8?" and write the other factors on the other side of the rectangle.

So, $12x - 8$ is equivalent to $4(3x - 2)$.

Glossary
expand
factor (an expression)

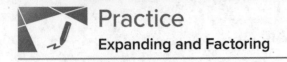

Practice
Expanding and Factoring

1. Respond to each of the following.

   a. Expand to write an equivalent expression: $\frac{-1}{4}(-8x + 12y)$.

   b. Factor to write an equivalent expression: $36a - 16$.

2. Lin missed math class on the day they worked on expanding and factoring. Kiran is helping Lin catch up.

   a. Lin understands that expanding is using the distributive property, but she doesn't understand what factoring is or why it works. How can Kiran explain factoring to Lin?

   b. Lin asks Kiran how the diagrams with boxes help with factoring. What should Kiran tell Lin about the boxes?

NAME _____ DATE _____ PERIOD _____

c. Lin asks Kiran to help her factor the expression -4xy − 12xz + 20xw. How can Kiran use this example to help Lin understand factoring?

3. Complete the equation with numbers that makes the expression on the right side of the equal sign equivalent to the expression on the left side.

   75a + 25b = _____ (_____ a + b)

4. Elena makes her favorite shade of purple paint by mixing 3 cups of blue paint, $1\frac{1}{2}$ cups of red paint, and $\frac{1}{2}$ of a cup of white paint. Elena has $\frac{2}{3}$ of a cup of white paint. **(Lesson 4-3)**

   a. Assuming she has enough red paint and blue paint, how much purple paint can Elena make?

   b. How much blue paint and red paint will Elena need to use with the $\frac{2}{3}$ of a cup of white paint?

5. Solve each equation. (Lesson 6-9)

a. $\frac{-1}{8}d - 4 = \frac{-3}{8}$

b. $\frac{-1}{4}m + 5 = 16$

c. $10b + -45 = -43$

d. $-8(y - 1.25) = 4$

e. $3.2(s + 10) = 32$

6. Select **all** the inequalities that have the same solutions as $-4x < 20$.
   (Lesson 6-13)

Ⓐ $-x < 5$

Ⓑ $4x > -20$

Ⓒ $4x < -20$

Ⓓ $x < -5$

Ⓔ $x > 5$

Ⓕ $x > -5$

Lesson 6-20

Combining Like Terms (Part 1)

NAME _____ DATE _____ PERIOD _____

Learning Goal Let's see how we can tell that expressions are equivalent.

 ## Warm Up
20.1 Why is it True?

Explain why each statement is true.

1. $5 + 2 + 3 = 5 + (2 + 3)$

2. $9a$ is equivalent to $11a - 2a$.

3. $7a + 4 - 2a$ is equivalent to $7a + -2a + 4$.

4. $8a - (8a - 8)$ is equivalent to 8.

Diego and Jada are both trying to write an expression with fewer terms that is equivalent to $7a + 5b - 3a + 4b$.

- Jada thinks $10a + 1b$ is equivalent to the original expression.

- Diego thinks $4a + 9b$ is equivalent to the original expression.

1. We can show expressions are equivalent by writing out all the variables. Explain why the expression on each row (after the first row) is equivalent to the expression on the row before it.

$$7a + 5b - 3a + 4b$$

$$(a + a + a + a + a + a + a) + (b + b + b + b + b) - (a + a + a) + (b + b + b + b)$$

$$(a + a + a + a) + (a + a + a) + (b + b + b + b + b) - (a + a + a) + (b + b + b + b)$$

$$(a + a + a + a) + (b + b + b + b + b) + (a + a + a) - (a + a + a) + (b + b + b + b)$$

$$(a + a + a + a) + (b + b + b + b + b) + (b + b + b + b)$$

$$(a + a + a + a) + (b + b + b + b + b + b + b + b + b)$$

$$4a + 9b$$

2. Here is another way we can rewrite the expressions. Explain why the expression on each row (after the first row) is equivalent to the expression on the row before it.

$$7a + 5b - 3a + 4b$$

$$7a + 5b + (-3a) + 4b$$

$$7a + (-3a) + 5b + 4b$$

$$(7 + -3)a + (5 + 4)b$$

$$4a + 9b$$

NAME _____ DATE _____ PERIOD _____

Are you ready for more?

Follow the instructions for a number puzzle.

- Take the number formed by the first 3 digits of your phone number and multiply it by 40.

- Add 1 to the result.

- Multiply by 500.

- Add the number formed by the last 4 digits of your phone number, and then add it again.

- Subtract 500.

- Multiply by $\frac{1}{2}$.

1. What is the final number?

2. How does this number puzzle work?

3. Can you invent a new number puzzle that gives a surprising result?

Activity

20.3 Making Sides Equal

Replace each ? with an expression that will make the left side of the equation equivalent to the right side.

Set A

1. $6x + ? = 10x$

2. $6x + ? = 2x$

3. $6x + ? = -10x$

4. $6x + ? = 0$

5. $6x + ? = 10$

Check your results with your partner and resolve any disagreements. Then move on to Set B.

Set B

1. $6x - ? = 2x$

2. $6x - ? = 10x$

3. $6x - ? = x$

4. $6x - ? = 6$

5. $6x - ? = 4x - 10$

Summary

Combining Like Terms (Part 1)

There are many ways to write equivalent expressions that may look very different from each other. We have several tools to find out if two expressions are equivalent.

- Two expressions are definitely not equivalent if they have different values when we substitute the same number for the variable. For example, $2(-3 + x) + 8$ and $2x + 5$ are not equivalent because when x is 1, the first expression equals 4 and the second expression equals 7.

- If two expressions are equal for many different values we substitute for the variable, then the expressions *may* be equivalent, but we don't know for sure. It is impossible to compare the two expressions for all values. To know for sure, we use properties of operations. For example, $2(-3 + x) + 8$ is equivalent to $2x + 2$ because:

$$2(-3 + x) + 8$$

$$-6 + 2x + 8 \qquad \text{by the distributive property}$$

$$2x + -6 + 8 \qquad \text{by the commutative property}$$

$$2x + (-6 + 8) \qquad \text{by the associative property}$$

$$2x + 2$$

NAME _____ DATE _____ PERIOD _____

 Practice
Combining Like Terms (Part 1)

1. Andre says that $10x + 6$ and $5x + 11$ are equivalent because they both equal 16 when x is 1. Do you agree with Andre? Explain your reasoning.

2. Select **all** expressions that can be subtracted from $9x$ to result in the expression $3x + 5$.

   (A.) $-5 + 6x$

   (B.) $5 - 6x$

   (C.) $6x + 5$

   (D.) $6x - 5$

   (E.) $-6x + 5$

3. Select **all** the statements that are true for any value of x.

   (A.) $7x + (2x + 7) = 9x + 7$

   (B.) $7x + (2x - 1) = 9x + 1$

   (C.) $\frac{1}{2}x + \left(3 - \frac{1}{2}x\right) = 3$

   (D.) $5x - (8 - 6x) = -x - 8$

   (E.) $0.4x - (0.2x + 8) = 0.2x - 8$

   (F.) $6x - (2x - 4) = 4x + 4$

4. For each situation, would you describe it with $x < 25$, $x > 25$, $x \leq 25$, or $x \geq 25$? (Lesson 6-13)

   a. The library is having a party for any student who read at least 25 books over the summer. Priya read x books and was invited to the party.

   b. Kiran read x books over the summer but was not invited to the party.

   c.

   d.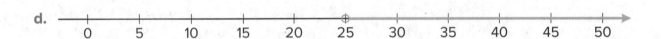

5. Consider the problem: A water bucket is being filled with water from a water faucet at a constant rate. When will the bucket be full? What information would you need to be able to solve the problem? (Lesson 2-9)

Lesson 6-21

Combining Like Terms (Part 2)

NAME _____ DATE _____ PERIOD _____

Learning Goal Let's see how to use properties correctly to write equivalent expressions.

 ## Warm Up
21.1 True or False?

Select **all** the statements that are true. Be prepared to explain your reasoning.

A. $4 - 2(3 + 7) = 4 - 2 \cdot 3 - 2 \cdot 7$

B. $4 - 2(3 + 7) = 4 + \text{-}2 \cdot 3 + \text{-}2 \cdot 7$

C. $4 - 2(3 + 7) = 4 - 2 \cdot 3 + 2 \cdot 7$

D. $4 - 2(3 + 7) = 4 - (2 \cdot 3 + 2 \cdot 7)$

Some students are trying to write an expression with fewer terms that is equivalent to $8 - 3(4 - 9x)$.

Noah says, "I worked the problem from left to right and ended up with $20 - 45x$."

$$8 - 3(4 - 9x)$$
$$5(4 - 9x)$$
$$20 - 45x$$

Lin says, "I started inside the parentheses and ended up with $23x$."

$$8 - 3(4 - 9x)$$
$$8 - 3(-5x)$$
$$8 + 15x$$
$$23x$$

Jada says, "I used the distributive property and ended up with $27x - 4$."

$$8 - 3(4 - 9x)$$
$$8 - (12 - 27x)$$
$$8 - 12 - (-27x)$$
$$27x - 4$$

Andre says, "I also used the distributive property, but I ended up with $-4 - 27x$."

$$8 - 3(4 - 9x)$$
$$8 - 12 - 27x$$
$$-4 - 27x$$

1. Do you agree with any of them? Explain your reasoning.

2. For each strategy that you disagree with, find and describe the errors.

NAME _____ DATE _____ PERIOD _____

Are you ready for more?

1. Jada's neighbor said, "My age is the difference between twice my age in 4 years and twice my age 4 years ago." How old is Jada's neighbor?

2. Another neighbor said, "My age is the difference between twice my age in 5 years and and twice my age 5 years ago." How old is this neighbor?

3. A third neighbor had the same claim for 17 years from now and 17 years ago, and a fourth for 21 years. Determine those neighbors' ages.

Activity

21.3 Grouping Differently

Diego was taking a math quiz. There was a question on the quiz that had the expression $8x - 9 - 12x + 5$. Diego's teacher told the class there was a typo and the expression was supposed to have one set of parentheses in it.

1. Where could you put parentheses in $8x - 9 - 12x + 5$ to make a new expression that is still equivalent to the original expression? How do you know that your new expression is equivalent?

2. Where could you put parentheses in $8x - 9 - 12x + 5$ to make a new expression that is not equivalent to the original expression? List as many different answers as you can.

Summary
Combining Like Terms (Part 2)

Combining like terms allows us to write expressions more simply with fewer terms. But it can sometimes be tricky with long expressions, parentheses, and negatives. It is helpful to think about some common errors that we can be aware of and try to avoid.

- $6x - x$ is not equivalent to 6.
  While it might be tempting to think that subtracting x makes the x disappear, the expression is really saying take $1 x$ away from $6 x$'s, and the distributive property tells us that $6x - x$ is equivalent to $(6 - 1)x$.

- $7 - 2x$ is not equivalent to $5x$.
  The expression $7 - 2x$ tells us to double an unknown amount and subtract it from 7. This is not always the same as taking 5 copies of the unknown.

- $7 - 4(x + 2)$ is not equivalent to $3(x + 2)$.
  The expression tells us to subtract 4 copies of an amount from 7, not to take $(7 - 4)$ copies of the amount.

If we think about the meaning and properties of operations when we take steps to rewrite expressions, we can be sure we are getting equivalent expressions and are not changing their value in the process.

NAME _____ DATE _____ PERIOD _____

Practice
Combining Like Terms (Part 2)

1. Consider each scenario.

- Noah says that $9x - 2x + 4x$ is equivalent to $3x$, because the subtraction sign tells us to subtract everything that comes after $9x$.

- Elena says that $9x - 2x + 4x$ is equivalent to $11x$, because the subtraction only applies to $2x$.

Do you agree with either of them? Explain your reasoning.

2. Identify the error in generating an expression equivalent to $4 + 2x - \frac{1}{2}(10 - 4x)$. Then correct the error.

$4 + 2x + \frac{-1}{2}(10 + \text{-}4x)$

$4 + 2x + \text{-}5 + 2x$

$4 + 2x - 5 + 2x$

$\text{-}1$

3. Select **all** expressions that are equivalent to $5x - 15 - 20x + 10$.

(A.) $5x - (15 + 20x) + 10$

(B.) $5x + \text{-}15 + \text{-}20x + 10$

(C.) $5(x - 3 - 4x + 2)$

(D.) $\text{-}5(\text{-}x + 3 + 4x + \text{-}2)$

(E.) $\text{-}15x - 5$

(F.) $\text{-}5(3x + 1)$

(G.) $\text{-}15\left(x - \frac{1}{3}\right)$

4. The school marching band has a budget of up to $750 to cover 15 new uniforms and competition fees that total $300. How much can they spend for one uniform? **(Lesson 6-14)**

    a. Write an inequality to represent this situation.

    b. Solve the inequality and describe what it means in the situation.

5. Solve the inequality that represents each story. Then interpret what the solution means in the story. **(Lesson 6-16)**

    a. For every $9 that Elena earns she gives x dollars to charity. This happens 7 times this month. Elena wants to be sure she keeps at least $42 from this month's earnings. $7(9 - x) \geq 42$

    b. Lin buys a candle that is 9 inches tall and burns down x inches per minute. She wants to let the candle burn for 7 minutes until it is less than 6 inches tall. $9 - 7x < 6$

6. A certain shade of blue paint is made by mixing $1\frac{1}{2}$ quarts of blue paint with 5 quarts of white paint. If you need a total of 16.25 gallons of this shade of blue paint, how much of each color should you mix? **(Lesson 4-3)**

Lesson 6-22

Combining Like Terms (Part 3)

NAME _____ DATE _____ PERIOD _____

Learning Goal Let's see how we can combine terms in an expression to write it with less terms.

 ## Warm Up
22.1 Are They Equal?

Select **all** expressions that are equal to $8 - 12 - (6 + 4)$.

- (A.) $8 - 6 - 12 + 4$
- (B.) $8 - 12 - 6 - 4$
- (C.) $8 - 12 + (6 + 4)$
- (D.) $8 - 12 - 6 + 4$
- (E.) $8 - 4 - 12 - 6$

 ## Activity
22.2 X's and Y's

Match each expression in Column A with an equivalent expression from Column B. Be prepared to explain your reasoning.

Column A	Column B
a. $(9x + 5y) + (3x + 7y)$	i. $12(x + y)$
b. $(9x + 5y) - (3x + 7y)$	ii. $12(x - y)$
c. $(9x + 5y) - (3x - 7y)$	iii. $6(x - 2y)$
d. $9x - 7y + 3x + 5y$	iv. $9x + 5y + 3x - 7y$
e. $9x - 7y + 3x - 5y$	v. $9x + 5y - 3x + 7y$
f. $9x - 7y - 3x - 5y$	vi. $9x - 3x + 5y - 7y$

Activity

22.3 Seeing Structure and Factoring

Write each expression with fewer terms. Show or explain your reasoning.

1. $3 \cdot 15 + 4 \cdot 15 - 5 \cdot 15$

2. $3x + 4x - 5x$

3. $3(x - 2) + 4(x - 2) - 5(x - 2)$

4. $3\left(\frac{5}{2}x + 6\frac{1}{2}\right) + 4\left(\frac{5}{2}x + 6\frac{1}{2}\right) - 5\left(\frac{5}{2}x + 6\frac{1}{2}\right)$

NAME _____ DATE _____ PERIOD _____

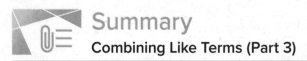

Summary
Combining Like Terms (Part 3)

Combining like terms is a useful strategy that we will see again and again in our future work with mathematical expressions. It is helpful to review the things we have learned about this important concept.

- Combining like terms is an application of the distributive property.
  For example:

  $$2x + 9x$$
  $$(2 + 9) \cdot x$$
  $$11x$$

- It often also involves the commutative and associative properties to change the order or grouping of addition.
  For example:

  $$2a + 3b + 4a + 5b$$
  $$2a + 4a + 3b + 5b$$
  $$(2a + 4a) + (3b + 5b)$$
  $$6a + 8b$$

- We can't change order or grouping when subtracting; so in order to apply the commutative or associative properties to expressions with subtraction, we need to rewrite subtraction as addition.
  For example:

  $$2a - 3b - 4a - 5b$$
  $$2a + \text{-}3b + \text{-}4a + \text{-}5b$$
  $$2a + \text{-}4a + \text{-}3b + \text{-}5b$$
  $$\text{-}2a + \text{-}8b$$
  $$\text{-}2a - 8b$$

- Since combining like terms uses properties of operations, it results in expressions that are equivalent.

- The like terms that are combined do not have to be a single number or variable; they may be longer expressions as well. Terms can be combined in any sum where there is a common factor in all the terms.
  For example, each term in the expression $5(x + 3) - 0.5(x + 3) + 2(x + 3)$ has a factor of $(x + 3)$. We can rewrite the expression with fewer terms by using the distributive property:

  $$5(x + 3) - 0.5(x + 3) + 2(x + 3)$$
  $$(5 - 0.5 + 2)(x + 3)$$
  $$6.5(x + 3)$$

Combining Like Terms (Part 3)

1. Jada says, "I can tell that $\frac{-2}{3}(x + 5) + 4(x + 5) - \frac{10}{3}(x + 5)$ equals 0 just by looking at it." Is Jada correct? Explain how you know.

2. In each row, decide whether the expression in Column A is equivalent to the expression in Column B. If they are not equivalent, show how to change one expression to make them equivalent.

Column A	Column B
a. $3x - 2x + 0.5x$	$1.5x$
b. $3(x + 4) - 2(x + 4)$	$x + 3$
c. $6(x + 4) - 2(x + 5)$	$2(2x + 7)$
d. $3(x + 4) - 2(x + 4) + 0.5(x + 4)$	1.5
e. $20\left(\frac{2}{5}x + \frac{3}{4}y - \frac{1}{2}\right)$	$\frac{1}{2}(16x + 30y - 20)$

NAME _____ DATE _____ PERIOD _____

3. For each situation, write an expression for the new balance using as few terms as possible. **(Lesson 6-20)**

   a. A checking account has a balance of -$126.89. A customer makes two deposits, one $3\frac{1}{2}$ times the other, and then withdraws $25.

   b. A checking account has a balance of $350. A customer makes two withdrawals, one $50 more than the other. Then he makes a deposit of $75.

4. Tyler is using the distributive property on the expression $9 - 4(5x - 6)$. Here is his work. **(Lesson 6-21)**

   $9 - 4(5x - 6)$
   $9 + (-4)(5x + -6)$
   $9 + -20x + -6$
   $3 - 20x$

   Mai thinks Tyler's answer is incorrect. She says, "If expressions are equivalent then they are equal for any value of the variable. Why don't you try to substitute the same value for x in all the equations and see where they are not equal?"

   a. Find the step where Tyler made an error.

b. Explain what he did wrong.

c. Correct Tyler's work.

5. Respond to each question. **(Lesson 6-13)**

    a. If $(11 + x)$ is positive, but $(4 + x)$ is negative, what is one number that x could be?

    b. If $(-3 + y)$ is positive, but $(-9 + y)$ is negative, what is one number that y could be?

    c. If $(-5 + z)$ is positive, but $(-6 + z)$ is negative, what is one number that z could be?

Lesson 6-23

Applications of Expressions

NAME _____ DATE _____ PERIOD _____

Learning Goal Let's use expressions to solve problems.

Warm Up
23.1 Algebra Talk: Equivalent to 0.75t − 21

Decide whether each expression is equivalent to $0.75t - 21$. Be prepared to explain how you know.

1. $\frac{3}{4}t - 21$ 2. $\frac{3}{4}(t - 21)$ 3. $0.75(t - 28)$ 4. $t - 0.25t - 21$

Activity
23.2 Two Ways to Calculate

Usually when you want to calculate something, there is more than one way to do it. For one or more of these situations, show how the two different ways of calculating are equivalent to each other.

1. Estimating the temperature in Fahrenheit when you know the temperature in Celsius:

   a. Double the temperature in Celsius, then add 30.

   b. Add 15 to the temperature in Celsius, then double the result.

2. Calculating a 15% tip on a restaurant bill:

   a. Take 10% of the bill amount, take 5% of the bill amount, and add those two values together.

b. Multiply the bill amount by 3, divide the result by 2, and then take $\frac{1}{10}$ of that result.

3. Changing a distance in miles to a distance in kilometers

   a. Take the number of miles, double it, then decrease the result by 20%.

   b. Divide the number of miles by 5, then multiply the result by 8.

Activity
23.3 Which Way?

You have two coupons to the same store: one for 20% off and one for $30 off. The cashier will let you use them both, and will let you decide in which order to use them.

- Mai says that it doesn't matter in which order you use them. You will get the same discount either way.

- Jada says that you should apply the 20% off coupon first, and then the $30 off coupon.

- Han says that you should apply the $30 off coupon first, and then the 20% off coupon.

- Kiran says that it depends on how much you are spending.

Do you agree with any of them? Explain your reasoning.

Learning Targets

Lesson	Learning Target(s)
6-1 Relationships between Quantities	• I can think of ways to solve some more complicated word problems.
6-2 Reasoning about Contexts with Tape Diagrams	• I can explain how a tape diagram represents parts of a situation and relationships between them. • I can use a tape diagram to find an unknown amount in a situation.
6-3 Reasoning about Equations with Tape Diagrams	• I can match equations and tape diagrams that represent the same situation. • If I have an equation, I can draw a tape diagram that shows the same relationship.

(continued on the next page)

(continued from the previous page)

Lesson	Learning Target(s)
6-4 Reasoning about Equations and Tape Diagrams (Part 1)	• I can draw a tape diagram to represent a situation where there is a known amount and several copies of an unknown amount and explain what the parts of the diagram represent. • I can find a solution to an equation by reasoning about a tape diagram or about what value would make the equation true.
6-5 Reasoning about Equations and Tape Diagrams (Part 2)	• I can draw a tape diagram to represent a situation where there is more than one copy of the same sum and explain what the parts of the diagram represent. • I can find a solution to an equation by reasoning about a tape diagram or about what value would make the equation true.
6-6 Distinguishing between Two Types of Situations	• I understand the similarities and differences between the two main types of equations we are studying in this unit. • When I have a situation or a tape diagram, I can represent it with an equation.
6-7 Reasoning about Solving Equations (Part 1)	• I can explain how a balanced hanger and an equation represent the same situation. • I can find an unknown weight on a hanger diagram and solve an equation that represents the diagram. • I can write an equation that describes the weights on a balanced hanger.

Lesson	Learning Target(s)
6-8 Reasoning about Solving Equations (Part 2)	• I can explain how a balanced hanger and an equation represent the same situation. • I can explain why some balanced hangers can be described by two different equations, one with parentheses and one without. • I can find an unknown weight on a hanger diagram and solve an equation that represents the diagram. • I can write an equation that describes the weights on a balanced hanger.
6-9 Dealing with Negative Numbers	• I can use the idea of doing the same to each side to solve equations that have negative numbers or solutions.
6-10 Different Options for Solving One Equation	• For an equation like $3(x + 2) = 15$, I can solve it in two different ways: by first dividing each side by 3, or by first rewriting $3(x + 2)$ using the distributive property. • For equations with more than one way to solve, I can choose the easier way depending on the numbers in the equation.
6-11 Using Equations to Solve Problems	• I can solve story problems by drawing and reasoning about a tape diagram or by writing and solving an equation.

(continued on the next page)

(continued from the previous page)

Lesson	Learning Target(s)
6-12 Solving Problems about Percent Increase or Decrease	• I can solve story problems about percent increase or decrease by drawing and reasoning about a tape diagram or by writing and solving an equation.
6-13 Reintroducing Inequalities	• I can explain what the symbols $\leq$ and $\geq$ mean. • I can represent an inequality on a number line. • I understand what it means for a number to make an inequality true.
6-14 Finding Solutions to Inequalities in Context	• I can describe the solutions to an inequality by solving a related equation and then reasoning about values that make the inequality true. • I can write an inequality to represent a situation.
6-15 Efficiently Solving Inequalities	• I can graph the solutions to an inequality on a number line. • I can solve inequalities by solving a related equation and then checking which values are solutions to the original inequality.

Lesson	Learning Target(s)
6-16 Interpreting Inequalities	• I can match an inequality to a situation it represents, solve it, and then explain what the solution means in the situation. • If I have a situation and an inequality that represents it, I can explain what the parts of the inequality mean in the situation.
6-17 Modeling with Inequalities	• I can use what I know about inequalities to solve real-world problems.
6-18 Subtraction in Equivalent Expressions	• I can organize my work when I use the distributive property. • I can re-write subtraction as adding the opposite and then rearrange terms in an expression.
6-19 Expanding and Factoring	• I can organize my work when I use the distributive property. • I can use the distributive property to rewrite expressions with positive and negative numbers. • I understand that factoring and expanding are words used to describe using the distributive property to write equivalent expressions.

(continued on the next page)

(continued from the previous page)

Lesson	Learning Target(s)
6-20 Combining Like Terms (Part 1)	• I can figure out whether two expressions are equivalent to each other. • When possible, I can write an equivalent expression that has fewer terms.
6-21 Combining Like Terms (Part 2)	• I am aware of some common pitfalls when writing equivalent expressions, and I can avoid them. • When possible, I can write an equivalent expression that has fewer terms.
6-22 Combining Like Terms (Part 3)	• Given an expression, I can use various strategies to write an equivalent expression. • When I look at an expression, I can notice if some parts have common factors and make the expression shorter by combining those parts.
6-23 Applications of Expressions	• I can write algebraic expressions to understand and justify a choice between two options.

Notes:

_____

_____

_____

Unit 7

Angles, Triangles, and Prisms

The Flatiron Building, in New York City, was constructed to fill a triangle-shaped property. At the end of this unit, you'll build your own triangular prism from scratch!

Topics
- Angle Relationships
- Drawing Polygons with Given Conditions
- Solid Geometry
- Let's Put It to Work

Unit 7

Angles, Triangles, and Prisms

Angle Relationships

Drawing Polygons with Given Conditions

Solid Geometry

Let's Put It To Work

Lesson 7-1

Relationships of Angles

NAME _____ DATE _____ PERIOD _____

Learning Goal Let's examine some special angles.

Warm Up
1.1 Visualizing Angles

1. Which angle is bigger?

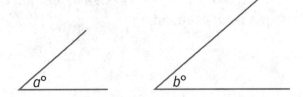

2. Identify an obtuse angle in the diagram.

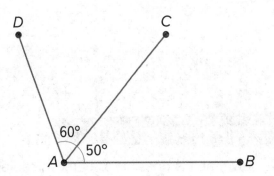

Activity
1.2 Pattern Block Angles

1. Trace one copy of every different pattern block. Each block contains either 1 or 2 angles with different degree measures. Which blocks have only 1 unique angle? Which have 2?

2. If you trace three copies of the hexagon so that one vertex from each hexagon touches the same point, as shown, they fit together without any gaps or overlaps. Use this to figure out the degree measure of the angle inside the hexagon pattern block.

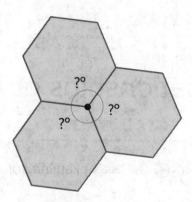

3. Figure out the degree measure of all of the other angles inside the pattern blocks that you traced in the first question. Be prepared to explain your reasoning.

Are you ready for more?

We saw that it is possible to fit three copies of a regular hexagon snugly around a point.

Each interior angle of a regular pentagon measures 108°. Is it possible to fit copies of a regular pentagon snugly around a point? If yes, how many copies does it take? If not, why not?

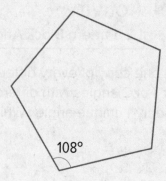

NAME _____ DATE _____ PERIOD _____

Activity
1.3 More Pattern Block Angles

1. Use pattern blocks to determine the measure of each of these angles.

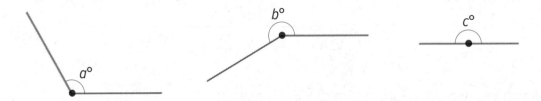

2. If an angle has a measure of 180°, then its sides form a straight line. An angle that forms a straight line is called a straight angle. Find as many different combinations of pattern blocks as you can that make a straight angle.

Activity
1.4 Measuring Like This or That

Tyler and Priya were both measuring angle *TUS*.

Priya thinks the angle measures 40 degrees. Tyler thinks the angle measures 140 degrees. Do you agree with either of them? Explain your reasoning.

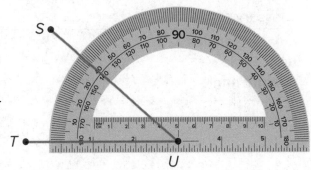

Summary
Relationships of Angles

When two lines intersect and form four equal angles, we call each one a **right angle**. A right angle measures 90°. You can think of a right angle as a quarter turn in one direction or the other.

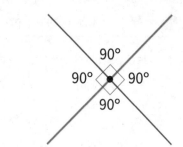

An angle in which the two sides form a straight line is called a **straight angle**. A straight angle measures 180°. A straight angle can be made by putting right angles together. You can think of a straight angle as a half turn, so that you are facing in the opposite direction after you are done.

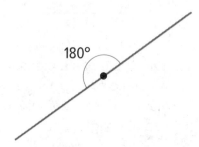

If you put two straight angles together, you get an angle that is 360°. You can think of this angle as turning all the way around so that you are facing the same direction as when you started the turn.

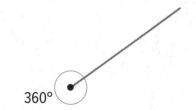

When two angles share a side and a vertex, and they don't overlap, we call them **adjacent angles**.

Glossary

adjacent angles
right angle
straight angle

NAME _____ DATE _____ PERIOD _____

Practice
Relationships of Angles

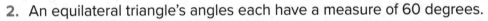

1. Here are questions about two types of angles.

   a. Draw a right angle. How do you know it's a right angle? What is its measure in degrees?

   b. Draw a straight angle. How do you know it's a straight angle? What is its measure in degrees?

2. An equilateral triangle's angles each have a measure of 60 degrees.

   a. Can you put copies of an equilateral triangle together to form a straight angle? Explain or show your reasoning.

   b. Can you put copies of an equilateral triangle together to form a right angle? Explain or show your reasoning.

3. Here is a square and some regular octagons.

   In this pattern, all of the angles inside the octagons have the same measure. The shape in the center is a square. Find the measure of one of the angles inside one of the octagons.

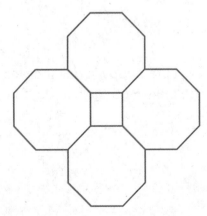

4. The height of the water in a tank decreases by 3.5 cm each day. When the tank is full, the water is 10 m deep. The water tank needs to be refilled when the water height drops below 4 m. (Lesson 6-17)

   a. Write a question that could be answered by solving the equation $10 - 0.035d = 4$.

   b. Is 100 a solution of $10 - 0.035d > 4$? Write a question that solving this problem could answer.

5. Use the distributive property to write an expression that is equivalent to each given expression. (Lesson 6-18)

   a. $-3(2x - 4)$

   b. $0.1(-90 + 50a)$

   c. $-7(-x - 9)$

   d. $\frac{4}{5}(10y + -x + -15)$

6. Lin's puppy is gaining weight at a rate of 0.125 pounds per day. Describe the weight gain in days per pound. (Lesson 2-3)

Lesson 7-2

Adjacent Angles

NAME _____ DATE _____ PERIOD _____

Learning Goal Let's look at some special pairs of angles.

Warm Up
2.1 Estimating Angle Measures

Estimate the degree measure of each indicated angle.

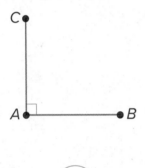 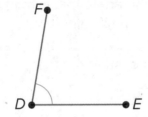 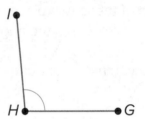

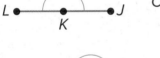 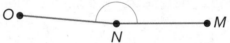 

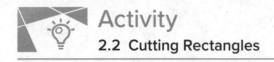

Activity

2.2 Cutting Rectangles

Your teacher will give you two small, rectangular papers.

1. On one of the papers, draw a small half-circle in the middle of one side.

2. Cut a straight line, starting from the center of the half-circle, all the way across the paper to make 2 separate pieces. (Your cut does not need to be perpendicular to the side of the paper.)

3. On each of these two pieces, measure the angle that is marked by part of a circle. Label the angle measure on the piece.

4. What do you notice about these angle measures?

5. Clare measured 70 degrees on one of her pieces. Predict the angle measure of her other piece.

6. On the other rectangular paper, draw a small quarter-circle in one of the corners.

7. Repeat the previous steps to cut, measure, and label the two angles marked by part of a circle.

8. What do you notice about these angle measures?

9. Priya measured 53 degrees on one of her pieces. Predict the angle measure of her other piece.

NAME _____ DATE _____ PERIOD _____

 Activity

2.3 Is It a Complement or Supplement?

1. Use the protractor in the picture to find the measure of angles *BCA* and *BCD*.

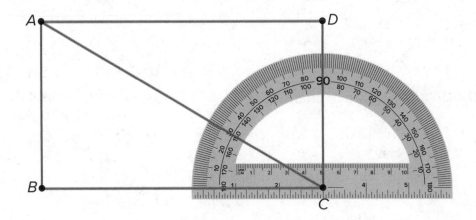

2. Explain how to find the measure of angle *ACD* without repositioning the protractor.

3. Use the protractor in the picture to find the measure of angles *LOK* and *LOM*.

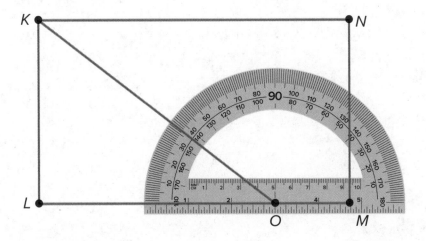

4. Explain how to find the measure of angle *KOM* without repositioning the protractor.

5. Angle *BAC* is a right angle. Find the measure of angle *CAD*.

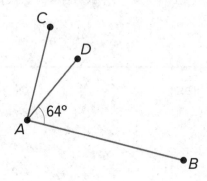

6. Point *O* is on line *RS*. Find the measure of angle *SOP*.

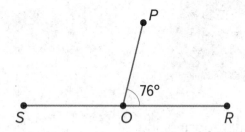

NAME _____ DATE _____ PERIOD _____

Are you ready for more?

Clare started with a rectangular piece of paper. She folded up one corner, and then folded up the other corner, as shown in the photos.

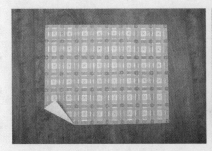  

1. Try this yourself with any rectangular paper. Fold the left corner up at any angle, and then fold the right corner up so that the edges of the paper meet.

2. Clare thought that the angle at the bottom looked like a 90 degree angle. Does yours also look like it is 90 degrees?

3. Can you explain why the bottom angle *always has to be* 90 degrees? Hint: the third photo shows Clare's paper, unfolded. The crease marks have dashed lines, and the line where the two paper edges met have a solid line. Mark these on your own paper as well.

If two angle measures add up to 90°, then we say the angles are **complementary**. Here are three examples of pairs of complementary angles.

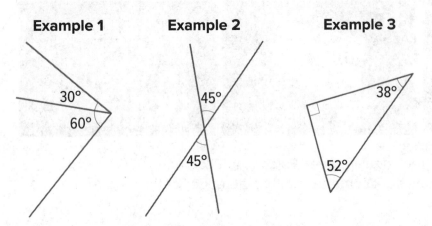

Example 1   **Example 2**   **Example 3**

30°
60°

45°
45°

38°
52°

If two angle measures add up to 180°, then we say the angles are **supplementary**. Here are three examples of pairs of supplementary angles.

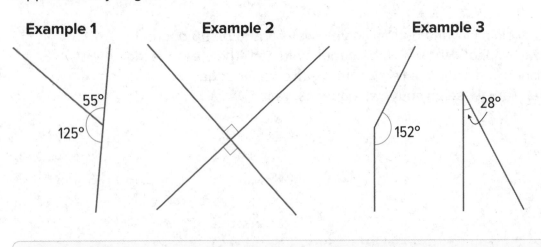

Example 1   **Example 2**   **Example 3**

55°
125°

152°

28°

Glossary

complementary
supplementary

NAME _____ DATE _____ PERIOD _____

 Practice
Adjacent Angles

1. Angles *A* and *C* are supplementary. Find the measure of angle *C*.

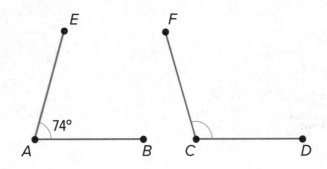

2. Respond to each of the following.

   a. List two pairs of angles in square *CDFG* that are
   complementary.

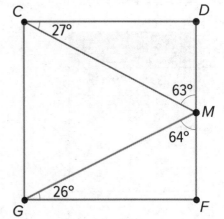

   b. Name three angles that sum to 180°.

3. Complete the equation with a number that makes the expression on the right side of the equal sign equivalent to the expression on the left side. (Lesson 6-22)

$$5x - 2.5 + 6x - 3 = \underline{\hspace{1.5cm}} (2x - 1)$$

4. Match each table with the equation that represents the same proportional relationship. (Lesson 2-4)

Tables

a.

x	y
2	8
3	12
4	16
5	20

b.

x	y
3	4.5
6	9
7	10.5
10	15

c.

x	y
2	$\frac{5}{2}$
4	5
6	$\frac{15}{2}$
12	15

Equations

i. $y = 1.5x$

ii. $y = 1.25x$

iii. $y = 4x$

Lesson 7-3

Nonadjacent Angles

NAME _____ DATE _____ PERIOD _____

Learning Goal Let's look at angles that are not right next to one another.

Warm Up
3.1 Finding Related Statements

Given a and b are numbers, and $a + b = 180$, which statements also must be true?

(A.) $a = 180 - b$

(B.) $a - 180 = b$

(C.) $360 = 2a + 2b$

(D.) $a = 90$ and $b = 90$

Activity
3.2 Polygon Angles

Use any useful tools in the geometry toolkit to identify any pairs of angles in these figures that are complementary or supplementary.

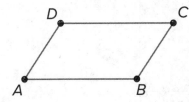
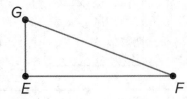

Activity

3.3 Vertical Angles

Use a straightedge to draw two intersecting lines. Use a protractor to measure all four angles whose vertex is located at the intersection.

Compare your drawing and measurements to those of the people in your group. Make a conjecture about the relationships between angle measures at an intersection.

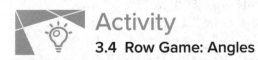

Activity

3.4 Row Game: Angles

Find the measure of the angles in one column. Your partner will work on the other column. Check in with your partner after you finish each row. Your answers in each row should be the same. If your answers aren't the same, work together to find the error and correct it.

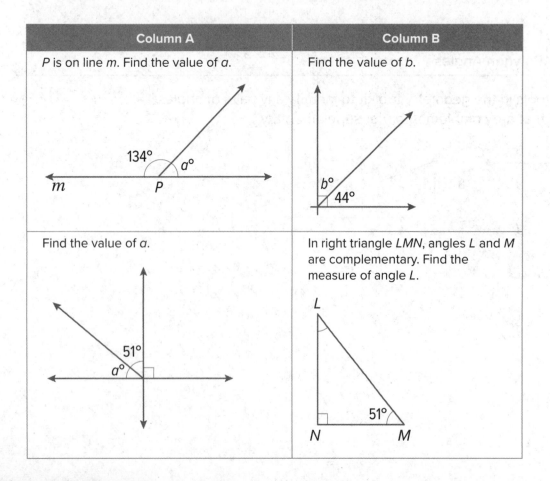

Column A	Column B
P is on line *m*. Find the value of *a*.	Find the value of *b*.
Find the value of *a*.	In right triangle *LMN*, angles *L* and *M* are complementary. Find the measure of angle *L*.

NAME _____ DATE _____ PERIOD _____

Column A	Column B
Angle C and angle E are supplementary. Find the measure of angle E. 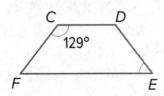	X is on line WY. Find the value of b. 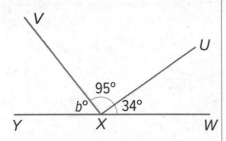
Find the value of c. 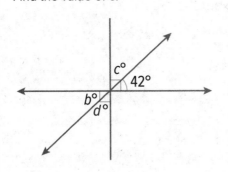	B is on line FW. Find the measure of angle CBW. 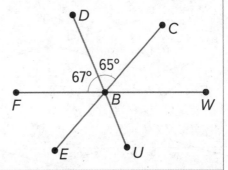
Two angles are complementary. One angle measures 37 degrees. Find the measure of the other angle.	Two angles are supplementary. One angle measures 127 degrees. Find the measure of the other angle.

When two lines cross, they form two pairs of **vertical angles**. Vertical angles are across the intersection point from each other.

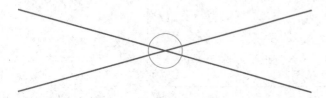

Vertical angles always have equal measure. We can see this because they are always supplementary with the same angle. For example:

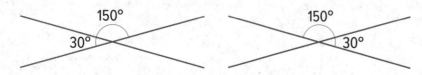

This is always true!

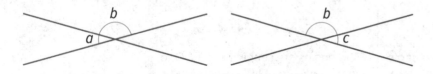

$a + b = 180$, so $a = 180 - b$.

$c + b = 180$, so $c = 180 - b$.

That means $a = c$.

Glossary
vertical angles

NAME _____ DATE _____ PERIOD _____

Practice
Nonadjacent Angles

1. Two lines intersect. Find the value of
   b and c.

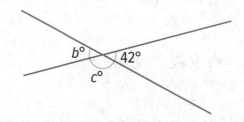

2. In this figure, angles *R* and *S* are complementary.
   Find the measure of angle *S*.

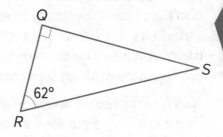

3. If two angles are both vertical and supplementary, can we determine
   the angles? Is it possible to be both vertical and complementary? If so,
   can you determine the angles? Explain how you know.

4. Match each expression in the first list with an equivalent expression
   from the second list. **(Lesson 6-22)**

   Expressions

   a. $5(x + 1) - 2x + 11$

   b. $2x + 2 + x + 5$

   c. $\frac{-3}{8}x - 9 + \frac{5}{8}x + 1$

   d. $2.06x - 5.53 + 4.98 - 9.02$

   e. $99x + 44$

   Equivalent Expressions

   i. $\frac{1}{4}x - 8$

   ii. $\frac{1}{2}(6x + 14)$

   iii. $11(9x + 4)$

   iv. $3x + 16$

   v. $2.06x + (\text{-}5.53) + 4.98 + (\text{-}9.02)$

5. Factor each expression. (Lesson 6-19)

   a. $15a - 13a$

   b. $-6x - 18y$

   c. $36abc + 54ad$

6. The directors of a dance show expect many students to participate but don't yet know how many students will come. The directors need 7 students to work on the technical crew. The rest of the students work on dance routines in groups of 9. For the show to work, they need at least 6 full groups working on dance routines. (Lesson 6-17)

   a. Write and solve an inequality to represent this situation and graph the solution on a number line.

   b. Write a sentence to the directors about the number of students they need.

7. A small dog gets fed $\frac{3}{4}$ cup of dog food twice a day. Using d for the number of days and f for the amount of food in cups, write an equation relating the variables. Use the equation to find how many days a large bag of dog food will last if it contains 210 cups of food. (Lesson 2-5)

Lesson 7-4

Solving for Unknown Angles

NAME _____ DATE _____ PERIOD _____

Learning Goal Let's figure out some missing angles.

 ## Warm Up
4.1 True or False: Length Relationships

Here are some line segments.

A   B   C   D
●————————●——————————————●————————●

Decide if each of these equations is true or false. Be prepared to explain your reasoning.

1. $CD + BC = BD$

2. $AB + BD = CD + AD$

3. $AC - AB = AB$

4. $BD - CD = AC - AB$

Activity

4.2 Info Gap: Angle Finding

Your teacher will give you either a *problem card* or a *data card*. Do not show or read your card to your partner.

If your teacher gives you the *problem card*:	If your teacher gives you the *data card*:
1. Silently read your card and think about what information you need to answer the question. 2. Ask your partner for the specific information that you need. 3. Explain to your partner how you are using the information to solve the problem. Continue to ask questions until you have enough information to solve the problem. 4. Share the *problem card* and solve the problem independently. 5. Read the *data card* and discuss your reasoning.	1. Silently read your card. 2. Ask your partner "*What specific information do you need*?" and wait for your partner to ask for information. If your partner asks for information that is not on the card, do not do the calculations for them. Tell them you don't have that information. 3. Before sharing the information, ask "*Why do you need that information?*" Listen to your partner's reasoning and ask clarifying questions. 4. Read the *problem card* and solve the problem independently. 5. Share the *data card* and discuss your reasoning.

Pause here so your teacher can review your work. Ask your teacher for a new set of cards and repeat the activity, trading roles with your partner.

NAME _____ DATE _____ PERIOD _____

 Activity
4.3 What's the Match?

Match each figure to an equation that represents what is seen in the figure. For each match, explain how you know they are a match.

Figure A

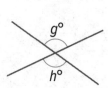

Figure B

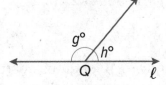

Figure C

Figure D

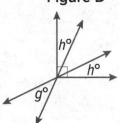

Figure E

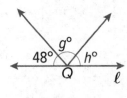

1. $g + h = 180$

2. $g = h$

3. $2h + g = 90$

4. $g + h + 48 = 180$

5. $g + h + 35 = 180$

1. What is the angle between the hour and minute hands of a clock at 3:00?

2. You might think that the angle between the hour and minute hands at 2:20 is 60 degrees, but it is not! The hour hand has moved beyond the 2. Calculate the angle between the clock hands at 2:20.

3. Find a time where the hour and minute hand are 40 degrees apart. (Assume that the time has a whole number of minutes.) Is there just one answer?

Summary
Solving for Unknown Angles

We can write equations that represent relationships between angles.

- The first pair of angles are supplementary, so $x + 42 = 180$.

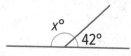

- The second pair of angles are vertical angles, so $y = 28$.

- Assuming the third pair of angles form a right angle, they are complementary, so $z + 64 = 90$.

NAME _____ DATE _____ PERIOD _____

Practice
Solving for Unknown Angles

1. *M* is a point on line segment *KL*. *NM* is a line segment. Select *all* the equations that represent the relationship between the measures of the angles in the figure.

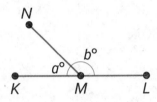

(A.) $a = b$

(B.) $a + b = 90$

(C.) $b = 90 - a$

(D.) $a + b = 180$

(E.) $180 - a = b$

(F.) $180 = b - a$

2. Which equation represents the relationship between the angles in the figure?

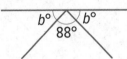

(A.) $88 + b = 90$

(B.) $88 + b = 180$

(C.) $2b + 88 = 90$

(D.) $2b + 88 = 180$

3. Segments *AB*, *EF*, and *CD* intersect at point *C*, and angle *ACD* is a right angle. Find the value of *g*.

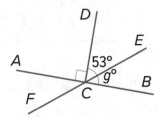

4. Select **all** the expressions that are the result of decreasing *x* by 80%. (Lesson 6-12)

(A.) $\frac{20}{100}x$

(B.) $x - \frac{80}{100}x$

(C.) $\frac{100 - 20}{100}x$

(D.) $0.80x$

(E.) $(1 - 0.8)x$

5. Andre is solving the equation $4\left(x + \frac{3}{2}\right) = 7$. He says, "I can subtract $\frac{3}{2}$ from each side to get $4x = \frac{11}{2}$ and then divide by 4 to get $x = \frac{11}{8}$." Kiran says, "I think you made a mistake." (Lesson 6-8)

a. How can Kiran know for sure that Andre's solution is incorrect?

NAME _____ DATE _____ PERIOD _____

b. Describe Andre's error and explain how to correct his work.

6. Solve each equation. (Lesson 6-7)

a. $\frac{1}{7}a + \frac{3}{4} = \frac{9}{8}$

b. $\frac{2}{3} + \frac{1}{5}b = \frac{5}{6}$

c. $\frac{3}{2} = \frac{4}{3}c + \frac{2}{3}$

d. $0.3d + 7.9 = 9.1$

e. $11.03 = 8.78 + 0.02e$

7. A train travels at a constant speed for a long distance. Write the two constants of proportionality for the relationship between distance traveled and elapsed time. Explain what each of them means. **(Lesson 2-5)**

Time Elapsed (hr)	Distance (mi)
1.2	54
3	135
4	180

Lesson 7-5

Using Equations to Solve for Unknown Angles

NAME _____ DATE _____ PERIOD _____

Learning Goal Let's figure out missing angles using equations.

 ## Warm Up
5.1 Is This Enough?

Tyler thinks that this figure has enough information to figure out the
values of *a* and *b*.

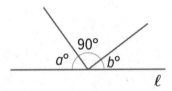

Do you agree? Explain your reasoning.

Elena and Diego each wrote equations to represent these diagrams. For each diagram, decide which equation you agree with, and solve it. You can assume that angles that look like right angles are indeed right angles.

1. Elena: $x = 35$

   Diego: $x + 35 = 180$

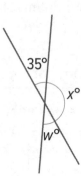

2. Elena: $35 + w + 41 = 180$

   Diego: $w + 35 = 180$

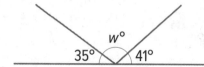

3. Elena: $w + 35 = 90$

   Diego: $2w + 35 = 90$

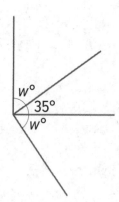

NAME _____ DATE _____ PERIOD _____

4. Elena: $2w + 35 = 90$

Diego: $w + 35 = 90$

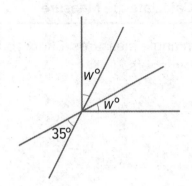

5. Elena: $w + 148 = 180$

Diego: $x + 90 = 148$

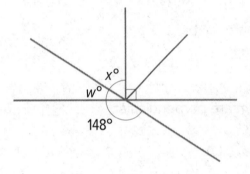

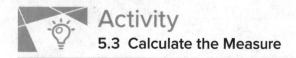
Find the unknown angle measures. Show your thinking. Organize it so it can be followed by others.

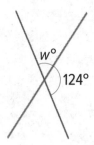

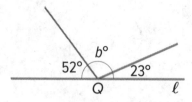

Lines ℓ and m are perpendicular.

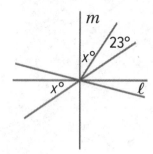

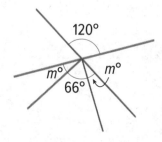

NAME _____ DATE _____ PERIOD _____

Are you ready for more?

The diagram contains three squares. Three additional segments have been drawn that connect corners of the squares. We want to find the exact value of $a + b + c$.

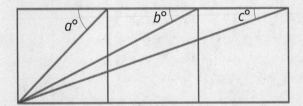

1. Use a protractor to measure the three angles. Use your measurements to conjecture about the value of $a + b + c$.

2. Find the exact value of $a + b + c$ by reasoning about the diagram.

 ## Summary

Using Equations to Solve for Unknown Angles

To find an unknown angle measure, sometimes it is helpful to write and solve an equation that represents the situation. For example, suppose we want to know the value of x in this diagram.

Using what we know about vertical angles, we can write the equation $3x + 90 = 144$ to represent this situation. Then we can solve the equation.

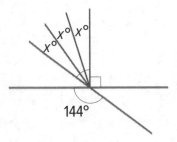

$$3x + 90 = 144$$

$$3x + 90 - 90 = 144 - 90$$

$$3x = 54$$

$$3x \cdot \frac{1}{3} = 54 \cdot \frac{1}{3}$$

$$x = 18$$

Practice

Using Equations to Solve for Unknown Angles

1. Segments *AB*, *DC*, and *EC* intersect at point *C*. Angle *DCE* measures 148°. Find the value of *x*.

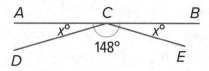

2. Line ℓ is perpendicular to line *m*. Find the value of *x* and *w*.

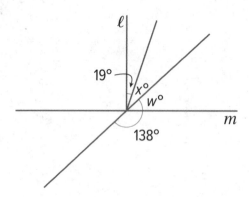

3. If you knew that two angles were complementary and were given the measure of one of those angles, would you be able to find the measure of the other angle? Explain your reasoning.

NAME _____ DATE _____ PERIOD _____

4. For each inequality, decide whether the solution is represented by $x < 4.5$ or $x > 4.5$. (Lesson 6-15)

   a. $-24 > -6(x - 0.5)$

   b. $-8x + 6 > -30$

   c. $-2(x + 3.2) < -15.4$

5. A runner ran $\frac{2}{3}$ of a 5-kilometer race in 21 minutes. They ran the entire race at a constant speed. (Lesson 4-2)

   a. How long did it take to run the entire race?

   b. How many minutes did it take to run 1 kilometer?

6. Jada, Elena, and Lin walked a total of 37 miles last week. Jada walked 4 more miles than Elena, and Lin walked 2 more miles than Jada. The diagram represents this situation: **(Lesson 6-12)**

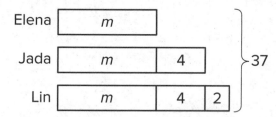

Find the number of miles that they each walked. Explain or show your reasoning.

7. Select **all** the expressions that are equivalent to $-36x + 54y - 90$.
   (Lesson 6-19)

   (A.) $-9(4x - 6y - 10)$

   (B.) $-18(2x - 3y + 5)$

   (C.) $-6(6x + 9y - 15)$

   (D.) $18(-2x + 3y - 5)$

   (E.) $-2(18x - 27y + 45)$

   (F.) $2(-18x + 54y - 90)$

Lesson 7-6

Building Polygons (Part 1)

NAME _____ DATE _____ PERIOD _____

Learning Goal Let's build shapes.

Warm Up
6.1 True or False: Signed Numbers

Decide whether each equation is true or false. Be prepared to explain your reasoning.

1. $4 \cdot (-6) = (-6) + (-6) + (-6) + (-6)$

2. $-8 \cdot 4 = (-8 \cdot 3) + 4$

3. $6 \cdot (-7) = 7 \cdot (-7) + 7$

4. $-10 - 6 = -10 - (-6)$

Activity
6.2 What Can You Build?

Your teacher will give you some strips of different lengths and fasteners you can use to attach the corners.

1. Use the pieces to build several polygons, including at least one triangle and one quadrilateral.

2. After you finish building several polygons, select one triangle and one quadrilateral that you have made.

   a. Measure all the angles in the two shapes you selected.

   b. Using these measurements along with the side lengths as marked, draw your triangle and quadrilateral as accurately as possible.

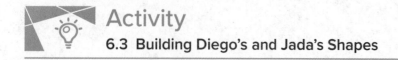

Activity
6.3 Building Diego's and Jada's Shapes

1. Diego built a quadrilateral using side lengths of 4 in, 5 in, 6 in, and 9 in.

   a. Build such a shape.

   b. Is your shape an identical copy of Diego's shape? Explain your reasoning.

2. Jada built a triangle using side lengths of 4 in, 5 in, and 8 in.

   a. Build such a shape.

   b. Is your shape an identical copy of Jada's shape? Explain your reasoning.

Activity
6.4 Building Han's Shape

Han built a polygon using side lengths of 3 in, 4 in, and 9 in.

1. Build such a shape.

2. What do you notice?

NAME _____ DATE _____ PERIOD _____

Summary
Building Polygons (Part 1)

Sometimes we are given a polygon and asked to find the lengths of the sides. What options do you have if you need to build a polygon with some side lengths?

Sometimes, we can make lots of different figures. For example, if you have side lengths 5, 7, 11, and 14, here are some of the many, many quadrilaterals we can make with those side lengths:

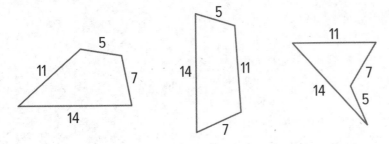

Sometimes, it is not possible to make a figure with certain side lengths. For example, 18, 1, 1, 1 (try it!).

We will continue to investigate the figures that can be made with given measures.

Practice
Building Polygons (Part 1)

1. A rectangle has side lengths of 6 units and 3 units. Could you make a quadrilateral that is not identical using the same four side lengths? If so, describe it.

2. Come up with an example of three side lengths that cannot possibly make a triangle and explain how you know.

3. Find x, y, and z. (Lesson 7-3)

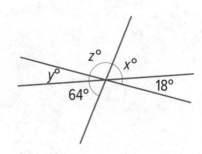

NAME _____ DATE _____ PERIOD _____

4. How many right angles need to be put together to make:

   a. 360 degrees?

   b. 180 degrees?

   c. 270 degrees?

   d. a straight angle?

5. Solve each equation. **(Lesson 6-8)**

   a. $\frac{1}{7}\left(x + \frac{3}{4}\right) = \frac{1}{8}$

   b. $\frac{9}{2} = \frac{3}{4}\left(z + \frac{2}{3}\right)$

   c. $1.5 = 0.6(w + 0.4)$

   d. $0.08(7.97 + v) = 0.832$

6. Respond to each of the following. (Lesson 4-3)

   a. You can buy 4 bottles of water from a vending machine for $7. At this rate, how many bottles of water can you buy for $28? If you get stuck, consider creating a table.

   b. It costs $20 to buy 5 sandwiches from a vending machine. At this rate, what is the cost for 8 sandwiches? If you get stuck, consider creating a table.

Lesson 7-7

Building Polygons (Part 2)

NAME _____ DATE _____ PERIOD _____

Learning Goal Let's build more triangles.

Warm Up
7.1 Where Is Lin?

At a park, the slide is 5 meters east of the swings. Lin is standing 3 meters away from the slide.

1. Draw a diagram of the situation including a place where Lin could be.

2. How far away from the swings is Lin in your diagram?

3. Where are some other places Lin could be?

Activity

7.2 How Long Is the Third Side?

Your teacher will give you some strips of different lengths and fasteners you can use to attach the corners.

1. Build as many different triangles as you can that have one side length of 5 inches and one of 4 inches. Record the side lengths of each triangle you build.

2. Are there any other lengths that could be used for the third side of the triangle but weren't in your set?

3. Are there any lengths that were in your set but could not be used as the third side of the triangle?

Are you ready for more?

Assuming you had access to strips of any length, and you used the 9-inch and 5-inch strips as the first two sides, complete the sentences:

1. The third side can't be _____ inches or longer.

2. The third side can't be _____ inches or shorter.

NAME _____ DATE _____ PERIOD _____

 ## Activity
7.3 Swinging the Sides Around

We'll explore a method for drawing a triangle that has three specific side lengths. Your teacher will give you a piece of paper showing a 4-inch segment as well as some instructions for which strips to use and how to connect them.

1. Follow these instructions to mark the possible endpoints of one side:

   a. Put your 4-inch strip directly on top of the 4-inch segment on the piece of paper. Hold it in place.

   b. For now, ignore the 3-inch strip on the left side. Rotate it so that it is out of the way.

   c. In the 3-inch strip on the *right* side, put the tip of your pencil in the hole on the end that is not connected to anything. Use the pencil to move the strip around its hinge, drawing all the places where a 3-inch side could end.

   d. Remove the connected strips from your paper.

2. What shape have you drawn while moving the 3-inch strip around? Why? Which tool in your geometry toolkit can do something similar?

3. Use your drawing to create two unique triangles, each with a base of length 4 inches and a side of length 3 inches. Use a different color to draw each triangle.

4. Reposition the strips on the paper so that the 4-inch strip is on top of the 4-inch segment again. In the 3-inch strip on the *left* side, put the tip of your pencil in the hole on the end that is not connected to anything. Use the pencil to move the strip around its hinge, drawing all the places where another 3-inch side could end.

5. Using a third color, draw a point where the two marks intersect. Using this third color, draw a triangle with side lengths of 4 inches, 3 inches, and 3 inches.

If we want to build a polygon with two given side lengths that share a vertex, we can think of them as being connected by a hinge that can be opened or closed.

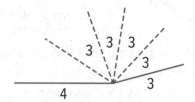

All of the possible positions of the endpoint of the moving side form a circle.

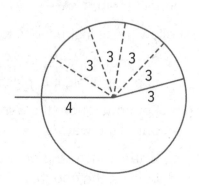

You may have noticed that sometimes it is not possible to build a polygon given a set of lengths. For example, if we have one really, really long segment and a bunch of short segments, we may not be able to connect them all. Here's what happens if you try to make a triangle with side lengths 21, 4, and 2.

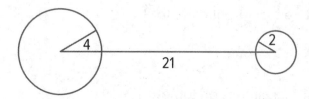

The short sides don't seem like they can meet up because they are too far away from each other.

If we draw circles of radius 4 and 2 on the endpoints of the side of length 21 to represent positions for the shorter sides, we can see that there are no places for the short sides that would allow them to meet up and form a triangle.

In general, the longest side length must be less than the sum of the other two side lengths. If not, we can't make a triangle!

If we *can* make a triangle with three given side lengths, it turns out that the measures of the corresponding angles will *always* be the same. For example, if two triangles have side lengths 3, 4, and 5, they will have the same corresponding angle measures.

NAME _____ DATE _____ PERIOD _____

Practice
Building Polygons (Part 2)

1. In the diagram, the length of segment *AB* is 10 units and the radius of the circle centered at *A* is 4 units. Use this to create two unique triangles, each with a side of length 10 and a side of length 4. Label the sides that have length 10 and 4.

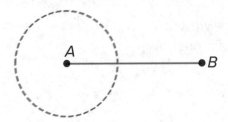

2. Select **all** the sets of three side lengths that will make a triangle.

Ⓐ 3, 4, 8

Ⓑ 7, 6, 12

Ⓒ 5, 11, 13

Ⓓ 4, 6, 12

Ⓔ 4, 6, 10

3. Based on signal strength, a person knows their lost phone is exactly 47 feet from the nearest cell tower. The person is currently standing 23 feet from the same cell tower. What is the closest the phone could be to the person? What is the farthest their phone could be from them?

4. Each row contains the degree measures of two complementary angles. Complete the table. (Lesson 7-2)

Measure of an Angle	Measure of Its Complement
80°	
25°	
54°	
x	

5. Here are two patterns made using identical rhombuses. Without using a protractor, determine the value of a and b. Explain or show your reasoning. (Lesson 7-1)

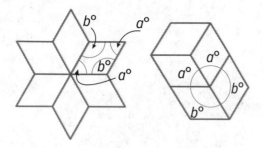

6. Mai's family is traveling in a car at a constant speed of 65 miles per hour. (Lesson 4-3)

a. At that speed, how long will it take them to travel 200 miles?

b. How far do they travel in 25 minutes?

Lesson 7-8

Triangles with Three Common Measures

NAME _____ DATE _____ PERIOD _____

Learning Goal Let's contrast triangles.

Warm Up
8.1 3 Sides; 3 Angles

Examine each set of triangles. What do you notice? What is the same about the triangles in the set? What is different?

Set 1:

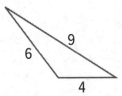
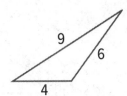
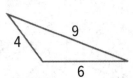

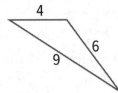
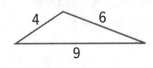
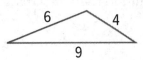

Set 2:

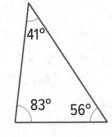
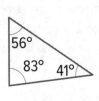

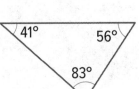
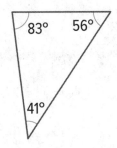

Activity

8.2 2 Sides and 1 Angle

Examine this set of triangles.

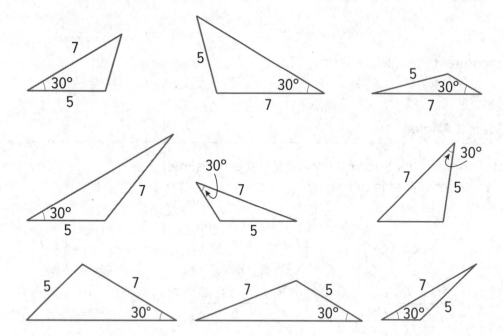

1. What is the same about the triangles in the set? What is different?

2. How many different triangles are there? Explain or show your reasoning.

NAME _____ DATE _____ PERIOD _____

Activity
8.3 2 Angles and 1 Side

Examine this set of triangles.

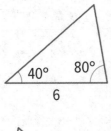
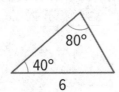
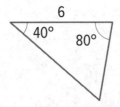

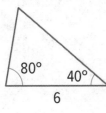
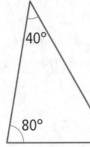
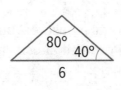

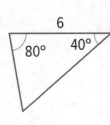
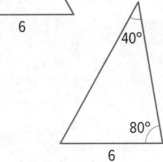

1. What is the same about the triangles in the set? What is different?

2. How many different triangles are there? Explain or show your reasoning.

Both of these quadrilaterals have a right angle and side lengths 4 and 5.

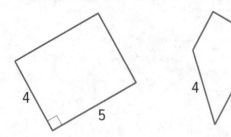

However, in one case, the right angle is *between* the two given side lengths; in the other, it is not.

If we create two triangles with three equal measures, but these measures are not next to each other in the same order, that usually means the triangles are different. Here is an example:

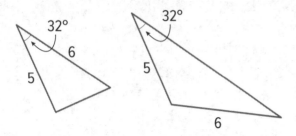

NAME _____ DATE _____ PERIOD _____

Practice
Triangles with Three Common Measures

1. Are these two triangles identical? Explain how you know.

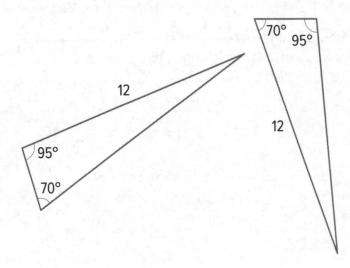

2. Are these triangles identical? Explain your reasoning.

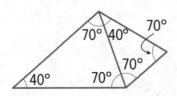

3. Tyler claims that if two triangles each have a side length of 11 units and a side length of 8 units, and also an angle measuring 100°, they must be identical to each other. Do you agree? Explain your reasoning.

4. The markings on the number line are equally spaced. Label the other markings on the number line.

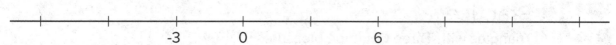

5. A passenger on a ship dropped his camera into the ocean. If it is descending at a rate of -4.2 meters per second, how long until it hits the bottom of the ocean, which is at -1,875 meters? **(Lesson 5-9)**

6. Apples cost $1.99 per pound. **(Lesson 4-3)**

   a. How much do $3\frac{1}{4}$ pounds of apples cost?

   b. How much do x pounds of apples cost?

   c. Clare spent $5.17 on apples. How many pounds of apples did Clare buy?

7. Diego has a glue stick with a diameter of 0.7 inches. He sets it down 3.5 inches away from the edge of the table, but it rolls onto the floor. How many rotations did the glue stick make before it fell off the table? **(Lesson 3-5)**

Lesson 7-9

Drawing Triangles (Part 1)

NAME _____ DATE _____ PERIOD _____

Learning Goal Let's see how many different triangles we can draw with certain measurements.

 ## Warm Up
9.1 Which One Doesn't Belong: Triangles

Which one doesn't belong?

Figure 1

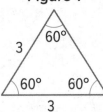

Figure 2

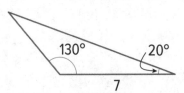

Figure 3

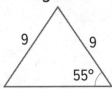

Figure 4

 ## Activity
9.2 Does Your Triangle Match Theirs?

Three students have each drawn a triangle. For each description:

• Draw a triangle with the given measurements.

• Measure and label the other side lengths and angle measures in your triangle.

• Decide whether the triangle you drew must be an identical copy of the triangle that the student drew. Explain your reasoning.

1. Jada's triangle has one angle measuring 75°.

2. Andre's triangle has one angle measuring 75° and one angle measuring 45°.

3. Lin's triangle has one angle measuring 75°, one angle measuring 45°, and one side measuring 5 cm.

 **Activity**

9.3 How Many Can You Draw?

1. Draw as many different triangles as you can with each of these sets of measurements:

   a. Two angles measure 60°, and one side measures 4 cm.

   b. Two angles measure 90°, and one side measures 4 cm.

   c. One angle measures 60°, one angle measures 90°, and one side measures 4 cm.

2. Which of these sets of measurements determine one unique triangle? Explain or show your reasoning.

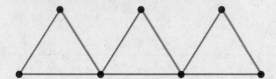

In the diagram, 9 toothpicks are used to make three equilateral triangles. Figure out a way to move only 3 of the toothpicks so that the diagram has exactly 5 equilateral triangles.

Summary
Drawing Triangles (Part 1)

Sometimes, we are given two different angle measures and a side length, and it is impossible to draw a triangle. For example, there is no triangle with side length 2 and angle measures 120° and 100°:

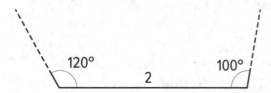

Sometimes, we are given two different angle measures and a side length between them, and we *can* draw a unique triangle. For example, if we draw a triangle with a side length of 4 between angles 90° and 60°, there is only one way they can meet up and complete to a triangle:

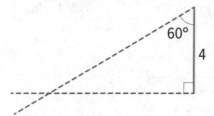

Any triangle drawn with these three conditions will be identical to the one above, with the same side lengths and same angle measures.

NAME _____ DATE _____ PERIOD _____

Practice
Drawing Triangles (Part 1)

1. Use a protractor to try to draw each triangle. Which of these three triangles is impossible to draw?

   (A.) A triangle where one angle measures 20° and another angle measures 45°

   (B.) A triangle where one angle measures 120° and another angle measures 50°

   (C.) A triangle where one angle measures 90° and another angle measures 100°

2. A triangle has an angle measuring 90°, an angle measuring 20°, and a side that is 6 units long. The 6-unit side is in between the 90° and 20° angles.

   a. Sketch this triangle and label your sketch with the given measures.

   b. How many unique triangles can you draw like this?

3. Respond to each of the following. **(Lesson 5-13)**

   a. Find a value for x that makes $-x$ less than $2x$.

   b. Find a value for x that makes $-x$ greater than $2x$.

4. One of the particles in atoms is called an electron. It has a charge of -1. Another particle in atoms is a proton. It has charge of +1.

   The overall charge of an atom is the sum of the charges of the electrons and the protons. Here is a list of common elements.

	Charge from Electrons	Charge from Protons	Overall Charge
Carbon	-6	+6	0
Aluminum	-10	+13	
Phosphide	-18	+15	
Iodide	-54	+53	
Tin	-50	+50	

   Find the overall charge for the rest of the atoms on the list. (Lesson 5-3)

5. A factory produces 3 bottles of sparkling water for every 7 bottles of plain water. If those are the only two products they produce, what percentage of their production is sparkling water? What percentage is plain? (Lesson 4-3)

Lesson 7-10

Drawing Triangles (Part 2)

NAME _____ DATE _____ PERIOD _____

Learning Goal Let's draw some more triangles.

 Warm Up
10.1 Using a Compass to Estimate Length

1. Draw a 40° angle.

2. Use a compass to make sure both sides of your angle have a length of 5 centimeters.

3. If you connect the ends of the sides you drew to make a triangle, is the third side longer or shorter than 5 centimeters? How can you use a compass to explain your answer?

Activity

10.2 Revisiting How Many Can You Draw?

1. Draw as many different triangles as you can with each of these sets of measurements:

    a. One angle measures 40°, one side measures 4 cm, and one side measures 5 cm.

    b. Two sides measure 6 cm, and one angle measures 100°.

2. Did either of these sets of measurements determine one unique triangle? How do you know?

NAME _____ DATE _____ PERIOD _____

Activity
10.3 Three Angles

1. Draw as many different triangles as you can with each of these sets of measurements:

   a. One angle measures 50°, one measures 60°, and one measures 70°.

   b. One angle measures 50°, one measures 60°, and one measures 100°.

2. Did either of these sets of measurements determine one unique triangle? How do you know?

Using *only* a compass and the edge of a blank index card, draw a perfectly equilateral triangle. (Note! The tools are part of the challenge! You may not use a protractor! You may not use a ruler!)

Summary

Drawing Triangles (Part 2)

A triangle has six measures: three side lengths and three angle measures.

If we are given three measures, then sometimes, there is no triangle that can be made. For example, there is no triangle with side lengths 1, 2, 5, and there is no triangle with all three angles measuring 150°.

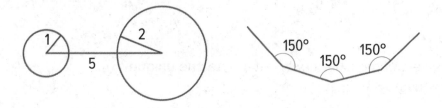

NAME _____ DATE _____ PERIOD _____

Sometimes, only one triangle can be made. By this we mean that any triangle we make will be the same, having the same six measures.

For example, if a triangle can be made with three given side lengths, then the corresponding angles will have the same measures.

Another example is shown here: an angle measuring 45° between two side lengths of 6 and 8 units. With this information, one unique triangle can be made.

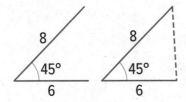

Sometimes, two or more different triangles can be made with three given measures. For example, here are two different triangles that can be made with an angle measuring 45° and side lengths 6 and 8. Notice the angle is not between the given sides.

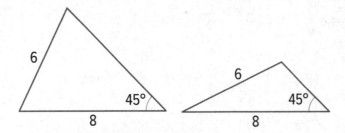

Three pieces of information about a triangle's side lengths and angle measures may determine no triangles, one unique triangle, or more than one triangle. It depends on the information.

1. A triangle has sides of length 7 cm, 4 cm, and 5 cm. How many unique triangles can be drawn that fit that description? Explain or show your reasoning.

2. A triangle has one side that is 5 units long and an adjacent angle that measures 25°. The two other angles in the triangle measure 90° and 65°. Complete the two diagrams to create two *different* triangles with these measurements.

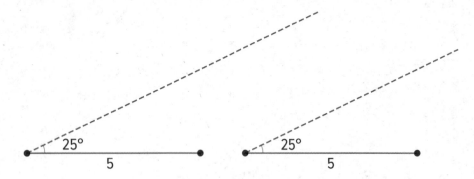

NAME _____ DATE _____ PERIOD _____

3. Is it possible to make a triangle that has angles measuring 90 degrees, 30 degrees, and 100 degrees? If so, draw an example. If not, explain your reasoning.

4. Segments *CD*, *AB*, and *FG* intersect at point *E*. Angle *FEC* is a right angle. Identify any pairs of angles that are complementary. **(Lesson 7-2)**

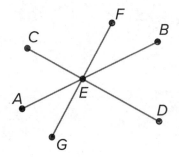

5. Match each equation to a step that will help solve the equation for x. (Lesson 5-15)

Equations

a. $3x = -4$

b. $-4.5 = x - 3$

c. $3 = \frac{-x}{3}$

d. $\frac{1}{3} = -3x$

e. $x - \frac{1}{3} = 0.4$

f. $3 + x = 8$

g. $\frac{x}{3} = 15$

h. $7 = \frac{1}{3} + x$

Steps

i. Add $\frac{1}{3}$ to each side.

ii. Add $\frac{-1}{3}$ to each side.

iii. Add 3 to each side.

iv. Add -3 to each side.

v. Multiply each side by 3.

vi. Multiply each side by -3.

vii. Multiply each side by $\frac{1}{3}$.

viii. Multiply each side by $\frac{-1}{3}$.

6. Respond to each question. (Lesson 4-8)

a. If you deposit $300 in an account with a 6% interest rate, how much will be in your account after 1 year?

b. If you leave this money in the account, how much will be in your account after 2 years?

Lesson 7-11

Slicing Solids

NAME _____ DATE _____ PERIOD _____

Learning Goal Let's see what shapes you get when you slice a three-dimensional object.

 ## Warm Up
11.1 Prisms, Pyramids, and Polyhedra

Describe each shape as precisely as you can.

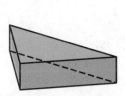 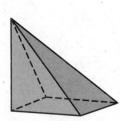 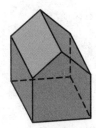

 ## Activity
11.2 What's the Cross Section?

Here are a rectangular prism and a pyramid with the same base and same height.

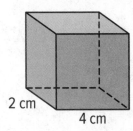 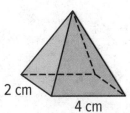

1. Think about slicing each solid parallel to its base, halfway up. What shape would each cross section be? What is the same about the two cross sections? What is different?

2. Think about slicing each solid parallel to its base, near the top. What shape would each cross section be? What is the same about the two cross sections? What is different?

Are you ready for more?

Describe the cross sections that would result from slicing each solid perpendicular to its base.

 Activity

11.3 Card Sort: Cross Sections

Your teacher will give you a set of cards. Sort the images into groups that make sense to you. Be prepared to explain your reasoning.

NAME _____ DATE _____ PERIOD _____

Activity
11.4 Drawing Cross Sections

Draw and describe each cross section.

1. Here is a picture of a rectangular prism, 4 units by 2 units by 3 units.

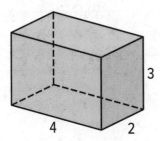

a. A plane cuts the prism parallel to the bottom and top faces.

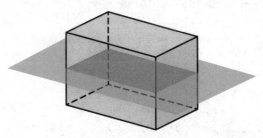

b. The plane moves up and cuts the prism at a different height.

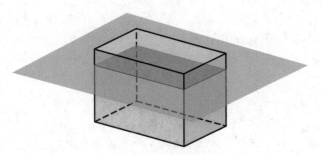

c. A vertical plane cuts the prism diagonally.

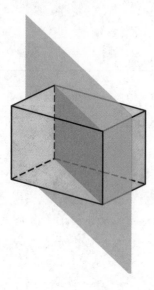

2. A square pyramid has a base that is 4 units by 4 units. Its height is also 4 units.

   a. A plane cuts the pyramid parallel to the base.

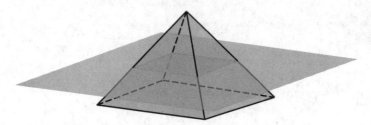

   b. A vertical plane cuts the pyramid.

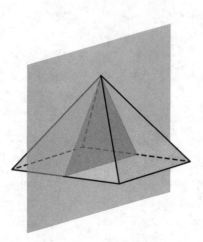

NAME _____ DATE _____ PERIOD _____

3. A cube has an edge of length 4.

   a. A plane cuts off the corner of the cube.

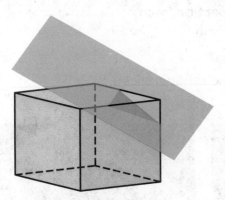

   b. The plane moves farther from the corner and makes a cut through the middle of the cube.

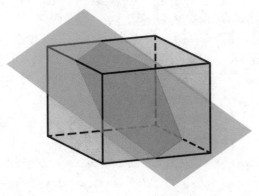

Summary
Slicing Solids

When we slice a three-dimensional object, we expose new faces that are two-dimensional. The two-dimensional face is a **cross section**. Many different cross sections are possible when slicing the same three-dimensional object.

Here are two peppers. One is sliced horizontally, and the other is sliced vertically, producing different cross sections.

 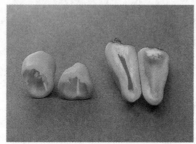 

The imprints of the slices represent the two-dimensional faces created by each slice.

It takes practice imagining what the cross section of a three-dimensional object will be for different slices. It helps to experiment and see for yourself what happens!

Glossary

base (of a prism or pyramid)
cross section
prism
pyramid

McGraw-Hill Education

NAME _____ DATE _____ PERIOD _____

Practice
Slicing Solids

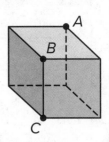

1. A cube is cut into two pieces by a single slice that passes through points *A*, *B*, and *C*. What shape is the cross section?

2. Describe how to slice the three-dimensional figure to result in each cross section.

Three-dimensional figure: Cross sections:

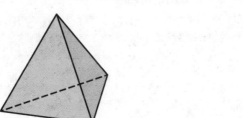

3. Here are two three-dimensional figures.

   Describe a way to slice one of the figures so that the cross section is a rectangle.

Figure A   **Figure B**

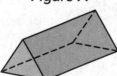

4. Each row contains the degree measures of two supplementary angles. Complete the table. (Lesson 7-2)

Measure of an Angle	Measure of Its Supplement
80°	
25°	
119°	
x	

5. Two months ago, the price, in dollars, of a cell phone was c. (Lesson 4-8)

a. Last month, the price of the phone increased by 10%. Write an expression for the price of the phone last month.

b. This month, the price of the phone decreased by 10%. Write an expression for the price of the phone this month.

c. Is the price of the phone this month the same as it was two months ago? Explain your reasoning.

Lesson 7-12

Volume of Right Prisms

NAME _____ DATE _____ PERIOD _____

Learning Goal Let's look at volumes of prisms.

Warm Up
12.1 Three Prisms with the Same Volume

Rectangles A, B, and C represent bases of three prisms.

Rectangle A   Rectangle B   Rectangle C

1. If each prism has the same height, which one will have the greatest volume, and which will have the least? Explain your reasoning.

2. If each prism has the same volume, which one will have the tallest height, and which will have the shortest? Explain your reasoning.

Activity

12.2 Finding Volume with Cubes

Your teacher will give you a paper with a shape on it and some snap cubes.

1. Using the face of a snap cube as your area unit, what is the area of the shape? Explain or show your reasoning.

2. Use snap cubes to build the shape from the paper. Add another layer of cubes on top of the shape you have built. Describe this three-dimensional object.

3. What is the volume of your object? Explain your reasoning.

4. Right now, your object has a height of 2. What would the volume be:

   a. if it had a height of 5?

   b. if it had a height of 8.5?

NAME _____ DATE _____ PERIOD _____

Activity

12.3 Can You Find the Volume?

Your teacher will give you a set of three-dimensional figures.

1. For each figure, determine whether the shape is a prism.

2. For each prism:

   a. Find the area of the base of the prism.

   b. Find the height of the prism.

   c. Calculate the volume of the prism.

	Is it a prism?	Area of Prism Base (cm²)	Height (cm)	Volume (cm³)
Figure A				
Figure B				
Figure C				
Figure D				
Figure E				
Figure F				

Are you ready for more?

Imagine a large, solid cube made out of 64 white snap cubes. Someone spray paints all 6 faces of the large cube blue. After the paint dries, they disassemble the large cube into a pile of 64 snap cubes.

1. How many of those 64 snap cubes have exactly 2 faces that are blue?

2. What are the other possible numbers of blue faces the cubes can have? How many of each are there?

3. Try this problem again with some larger-sized cubes that use more than 64 snap cubes to build. What patterns do you notice?

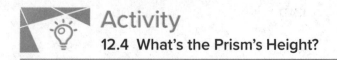

Activity

12.4 What's the Prism's Height?

There are 4 different prisms that all have the same volume. Here is what the base of each prism looks like.

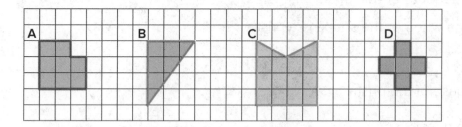

1. Order the prisms from shortest to tallest. Explain your reasoning.

2. If the volume of each prism is 60 units^3, what would be the height of each prism?

3. For a volume other than 60 units^3, what could be the height of each prism?

4. Discuss your thinking with your partner. If you disagree, work to reach an agreement.

NAME _____ DATE _____ PERIOD _____

Summary
Volume of Right Prisms

Any cross section of a prism that is parallel to the base will be identical to the base. This means we can slice prisms up to help find their volume.

For example, if we have a rectangular prism that is 3 units tall and has a base that is 4 units by 5 units, we can think of this as 3 layers, where each layer has 4 • 5 cubic units.

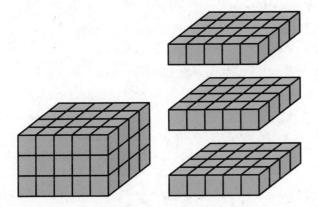

That means the volume of the original rectangular prism is 3(4 • 5) cubic units.

This works with any prism! If we have a prism with height 3 cm that has a base of area 20 cm², then the volume is 3 • 20 cm³ regardless of the shape of the base.

In general, the volume of a prism with height h and area B is $V = B \cdot h$.

For example, these two prisms both have a volume of 100 cm³.

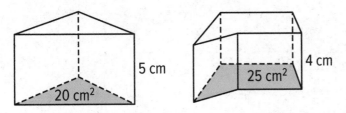

5 cm

20 cm²

4 cm

25 cm²

Glossary

volume

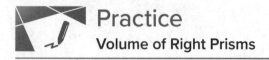

Practice
Volume of Right Prisms

1. For each prism, shade one of its bases.

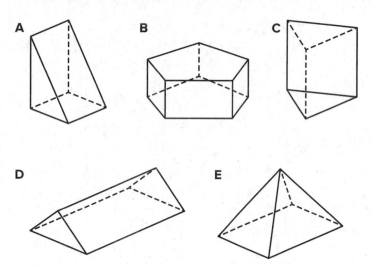

2. The volume of both of these trapezoidal prisms is 24 cubic units. Their heights are 6 and 8 units, as labeled. What is the area of a trapezoidal base of each prism?

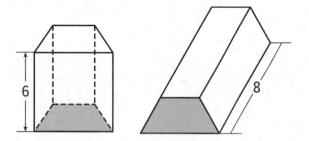

NAME _____ DATE _____ PERIOD _____

3. Two angles are complementary. One has a measure of 19 degrees. What is the measure of the other? (Lesson 7-2)

4. Two angles are supplementary. One has a measure that is twice as large as the other. Find the two angle measures. (Lesson 7-2)

5. Match each expression in the first list with an equivalent expression from the second list. (Lesson 6-22)

Expressions

a. $7(x + 2) - x + 3$

b. $6x + 3 + 4x + 5$

c. $\frac{-2}{5}x - 7 + \frac{3}{5}x - 3$

d. $8x - 5 + 4 - 9$

e. $24x + 36$

Equivalent Expressions

i. $\frac{1}{5}x - 10$

ii. $6x + 17$

iii. $2(5x + 4)$

iv. $12(2x + 3)$

v. $8x + (-5) + 4 + (-9)$

6. Clare paid 50% more for her notebook than Priya paid for hers. Priya paid x for her notebook and Clare paid y dollars for hers. Write an equation that represents the relationship between y and x. (Lesson 4-8)

Lesson 7-13

Decomposing Bases for Area

NAME _____ DATE _____ PERIOD _____

Learning Goal Let's look at how some people use volume.

Warm Up
13.1 Are These Prisms?

1. Which of these solids are prisms? Explain how you know.

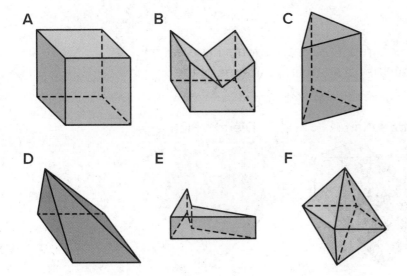

2. For each of the prisms, what does the base look like?

   a. Shade one base in the picture.

   b. Draw a cross section of the prism parallel to the base.

A box of chocolates is a prism with a base in the shape of a heart and a height of 2 inches. Here are the measurements of the base.

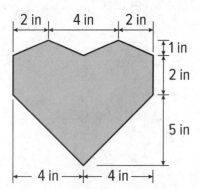

To calculate the volume of the box, three different students have each drawn line segments showing how they plan to find the area of the heart-shaped base.

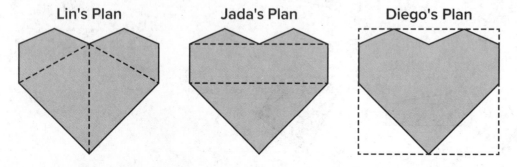

Lin's Plan   Jada's Plan   Diego's Plan

1. For each student's plan, describe the shapes the student must find the area of and the operations they must use to calculate the total area.

NAME _____ DATE _____ PERIOD _____

2. Although all three methods could work, one of them requires measurements that are not provided. Which one is it?

3. Between you and your partner, decide which of you will use which of the remaining two methods.

4. Using the quadrilaterals and triangles drawn in your selected plan, find the area of the base.

5. Trade with a partner and check each other's work. If you disagree, work to reach an agreement.

6. Return their work. Calculate the volume of the box of chocolates.

Are you ready for more?

The box has 30 pieces of chocolate in it, each with a volume of 1 in^3. If all the chocolates melt into a solid layer across the bottom of the box, what will be the height of the layer?

A house-shaped prism is created by attaching a triangular prism on top of a rectangular prism.

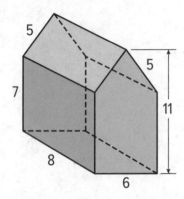

1. Draw the base of this prism and label its dimensions.

2. What is the area of the base? Explain or show your reasoning.

3. What is the volume of the prism?

NAME _____ DATE _____ PERIOD _____

Summary
Decomposing Bases for Area

To find the area of any polygon, you can decompose it into rectangles and triangles. There are always many ways to decompose a polygon.

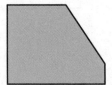 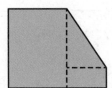 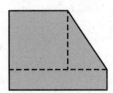 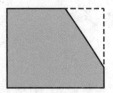

Sometimes it is easier to enclose a polygon in a rectangle and subtract the area of the extra pieces.

To find the volume of a prism with a polygon for a base, you find the area of the base, *B*, and multiply by the height, *h*.

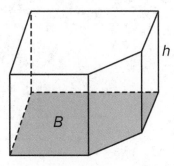

$$V = Bh$$

Practice
Decomposing Bases for Area

1. You find a crystal in the shape of a prism. Find the volume of the crystal.

   The point B is directly underneath point E, and the following lengths are known:

   - From A to B: 2 mm
   - From B to C: 3 mm
   - From A to F: 6 mm
   - From B to E: 10 mm
   - From C to D: 7 mm
   - From A to G: 4 mm

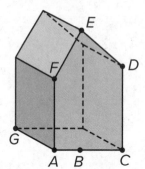

2. A rectangular prism with dimensions 5 inches by 13 inches by 10 inches was cut to leave a piece as shown in the image. What is the volume of this piece? What is the volume of the other piece not pictured?

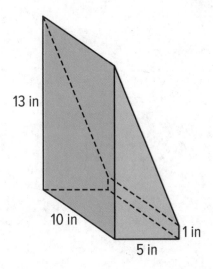

13 in

10 in

1 in

5 in

NAME _____ DATE _____ PERIOD _____

3. A triangle has one side that is 7 cm long and another side that is
3 cm long. (Lesson 7-9)

   a. Sketch this triangle and label your sketch with the given measures.
(If you are stuck, try using a compass or cutting some straws to these
two lengths.)

   b. Draw one more triangle with these measures that is not identical to
your first triangle.

   c. Explain how you can tell they are not identical.

4. Select **all** equations that represent a relationship between angles in the figure. (Lesson 7-4)

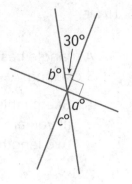

(A.) $90 - 30 = b$

(B.) $30 + b = a + c$

(C.) $a + c + 30 + b = 180$

(D.) $a = 30$

(E.) $a = c = 30$

(F.) $90 + a + c = 180$

5. A mixture of punch contains 1 quart of lemonade, 2 cups of grape juice, 4 tablespoons of honey, and $\frac{1}{2}$ gallon of sparkling water. Find the percentage of the punch mixture that comes from each ingredient. Round your answers to the nearest tenth of a percent. (Hint: 1 cup = 16 tablespoons) (Lesson 4-9)

Lesson 7-14

Surface Area of Right Prisms

NAME _____ DATE _____ PERIOD _____

Learning Goal Let's look at the surface area of prisms.

 Warm Up
14.1 Multifaceted

Your teacher will show you a prism.

1. What are some things you could measure about the object?

2. What units would you use for these measurements?

Activity

14.2 So Many Faces

Here is a picture of your teacher's prism:

Three students are trying to calculate the surface area of this prism.

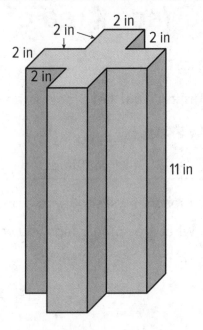

- Noah says, "This is going to be a lot of work. We have to find the areas of 14 different faces and add them up."

- Elena says, "It's not so bad. All 12 rectangles are identical copies, so we can find the area for one of them, multiply that by 12 and then add on the areas of the 2 bases."

- Andre says, "Wait, I see another way! Imagine unfolding the prism into a net. We can use 1 large rectangle instead of 12 smaller ones."

1. Do you agree with any of them? Explain your reasoning.

2. How big is the "1 large rectangle" Andre is talking about? Explain or show your reasoning. If you get stuck, consider drawing a net for the prism.

NAME _____ DATE _____ PERIOD _____

3. Will Noah's method always work for finding the surface area of any prism? Elena's method? Andre's method? Be prepared to explain your reasoning.

4. Which method do you prefer? Why?

Activity

14.3 Revisiting a Pentagonal Prism

1. Between you and your partner, choose who will use each of these two methods to find the surface area of the prism.

- adding the areas of all the faces

- using the perimeter of the base

2. Use your chosen method to calculate the surface area of the prism. Show your thinking. Organize it so it can be followed by others.

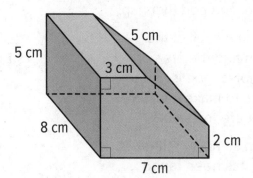

3. Trade papers with your partner and check their work. Discuss your thinking. If you disagree, work to reach an agreement.

In a deck of cards, each card measures 6 cm by 9 cm.

1. When stacked, the deck is 2 cm tall, as shown in the first photo. Find the volume of this deck of cards.

2. Then the cards are fanned out, as shown in the second picture. The distance from the rightmost point on the bottom card to the rightmost point on the top card is now 7 cm instead of 2 cm. Find the volume of the new stack.

Summary
Surface Area of Right Prisms

To find the surface area of a three-dimensional figure whose faces are made up of polygons, we can find the area of each face, and add them up!

Sometimes there are ways to simplify our work. For example, all of the faces of a cube with side length s are the same. We can find the area of one face, and multiply by 6. Since the area of one face of a cube is s^2, the surface area of a cube is $6s^2$. We can use this technique to make it faster to find the surface area of any figure that has faces that are the same.

For prisms, there is another way. We can treat the prism as having three parts: two identical bases, and one long rectangle that has been taped along the edges of the bases. The rectangle has the same height as the prism, and its width is the perimeter of the base. To find the surface area, add the area of this rectangle to the areas of the two bases.

> **Glossary**
> _____
> **surface area**

NAME _____ DATE _____ PERIOD _____

Practice
Surface Area of Right Prisms

1. Edge lengths are given in units. Find the surface area of each prism in square units.

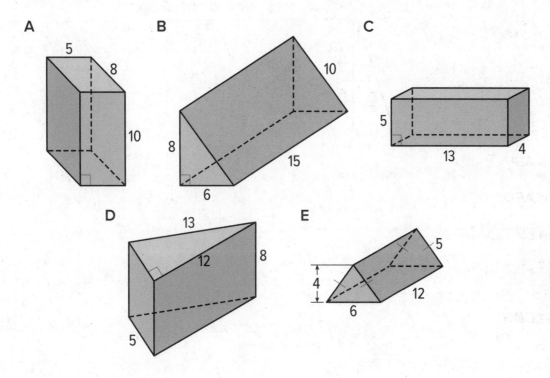

2. Priya says, "No matter which way you slice this rectangular prism, the cross section will be a rectangle." Mai says, "I'm not so sure." Describe a slice that Mai might be thinking of. (Lesson 7-11)

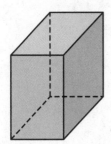

3. *B* is the intersection of line *AC* and line *ED*. Find the measure of each of the angles. (Lesson 7-5)

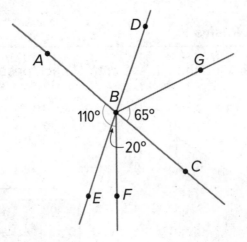

a. Angle *ABF*

b. Angle *ABD*

c. Angle *EBC*

d. Angle *FBC*

e. Angle *DBG*

4. Write each expression with fewer terms. (Lesson 6-20)

   a. $12m - 4m$

   b. $12m - 5k + m$

   c. $9m + k - (3m - 2k)$

5. Respond to each of the following. (Lesson 4-9)

   a. Find 44% of 625 using the facts that 40% of 625 is 250 and 4% of 625 is 25.

   b. What is 4.4% of 625?

   c. What is 0.44% of 625?

Lesson 7-15

Distinguishing Volume and Surface Area

NAME _____ DATE _____ PERIOD _____

Learning Goal Let's work with surface area and volume in context.

Warm Up
15.1 The Science Fair

Mai's science teacher told her that when there is more ice touching the water in a glass, the ice melts faster. She wants to test this statement so she designs her science fair project to determine if crushed ice or ice cubes will melt faster in a drink.

She begins with two cups of warm water. In one cup, she puts a cube of ice. In a second cup, she puts crushed ice with the same volume as the cube. What is your hypothesis? Will the ice cube or crushed ice melt faster, or will they melt at the same rate? Explain your reasoning.

Activity
15.2 Revisiting the Box of Chocolates

The other day, you calculated the volume of this heart-shaped box of chocolates.

The depth of the box is 2 inches. How much cardboard is needed to create the box?

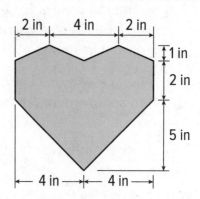

Your teacher will give you cards with different figures and questions on them.

1. Sort the cards into two groups based on whether it would make more sense to think about the surface area or the volume of the figure when answering the question. Pause here so your teacher can review your work.

2. Your teacher will assign you a card to examine more closely. What additional information would you need to be able to answer the question on your card?

3. Estimate reasonable measurements for the figure on your card.

4. Use your estimated measurements to calculate the answer to the question.

Are you ready for more?

A cake is shaped like a square prism. The top is 20 centimeters on each side, and the cake is 10 centimeters tall. It has frosting on the sides and on the top, and a single candle on the top at the exact center of the square. You have a knife and a 20-centimeter ruler.

1. Find a way to cut the cake into 4 fair portions, so that all 4 portions have the same amount of cake and frosting.

2. Find another way to cut the cake into 4 fair portions.

3. Find a way to cut the cake into 5 fair portions.

Activity

15.4 A Wheelbarrow of Concrete

A wheelbarrow is being used to carry wet concrete. Here are its dimensions.

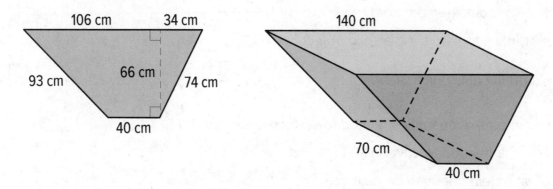

1. What volume of concrete would it take to fill the tray?

2. After dumping the wet concrete, you notice that a thin film is left on the inside of the tray. What is the area of the concrete coating the tray? (Remember, there is no top.)

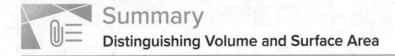

Summary
Distinguishing Volume and Surface Area

Sometimes we need to find the volume of a prism, and sometimes we need to find the surface area.

Here are some examples of quantities related to volume:

- how much water a container can hold
- how much material it took to build a solid object

Volume is measured in cubic units, like in^3 or m^3.

Here are some examples of quantities related to surface area:

- how much fabric is needed to cover a surface
- how much of an object needs to be painted

Surface area is measured in square units, like in^2 or m^2.

NAME _____ DATE _____ PERIOD _____

Practice
Distinguishing Volume and Surface Area

1. Here is the base of a prism.

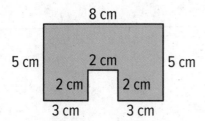

8 cm

5 cm   2 cm   5 cm

2 cm   2 cm

3 cm   3 cm

    a. If the height of the prism is 5 cm, what is its surface area? What is its volume?

    b. If the height of the prism is 10 cm, what is its surface area? What is its volume?

    c. When the height doubled, what was the percent increase for the surface area? For the volume?

2. Select **all** the situations where knowing the volume of an object would be more useful than knowing its surface area.

    A. determining the amount of paint needed to paint a barn

    B. determining the monetary value of a piece of gold jewelry

    C. filling an aquarium with buckets of water

    D. deciding how much wrapping paper a gift will need

    E. packing a box with watermelons for shipping

    F. charging a company for ad space on your race car

    G. measuring the amount of gasoline left in the tank of a tractor

3. Han draws a triangle with a 50° angle, a 40° angle, and a side of length 4 cm as shown. Can you draw a different triangle with the same conditions? (Lesson 7-9)

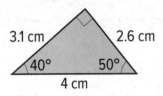

3.1 cm   2.6 cm
40°   50°
4 cm

4. Angle H is half as large as angle J. Angle J is one fourth as large as angle K. Angle K has measure 240 degrees. What is the measure of angle H? (Lesson 7-3)

5. The Colorado state flag consists of three horizontal stripes of equal height. The side lengths of the flag are in the ratio 2 : 3. The diameter of the gold-colored disk is equal to the height of the center stripe. What percentage of the flag is gold? (Lesson 4-9)

Lesson 7-16

Applying Volume and Surface Area

NAME _____ DATE _____ PERIOD _____

Learning Goal Let's explore things that are proportional to volume or surface area.

 **Warm Up**

16.1 You Decide

For each situation, decide if it requires Noah to calculate surface area or volume. Explain your reasoning.

1. Noah is planning to paint the bird house he built. He is unsure if he has enough paint.

2. Noah is planning to use a box with a trapezoid base to hold modeling clay. He is unsure if the clay will all fit in the box.

At a daycare, Kiran sees children climbing on this foam play structure.

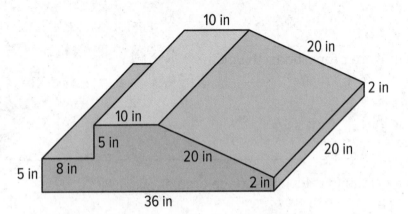

Kiran is thinking about building a structure like this for his younger cousins to play on.

1. The entire structure is made out of soft foam so the children don't hurt themselves. How much foam would Kiran need to build this play structure?

2. The entire structure is covered with vinyl so it is easy to wipe clean. How much vinyl would Kiran need to build this play structure?

3. The foam costs 0.8¢ per in³. Here is a table that lists the costs for different amounts of vinyl. What is the total cost for all the foam and vinyl needed to build this play structure?

Vinyl (in²)	Cost ($)
75	0.45
125	0.75

NAME _____ DATE _____ PERIOD _____

Are you ready for more?

When he examines the play structure more closely, Kiran realizes it is really two separate pieces that are next to each other.

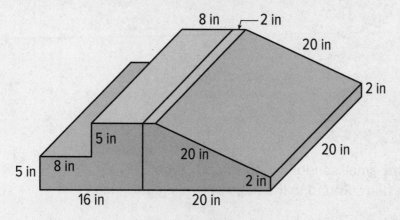

1. How does this affect the amount of foam in the play structure?

2. How does this affect the amount of vinyl covering the play structure?

Activity

16.3 Filling the Sandbox

The daycare has two sandboxes that are both prisms with regular hexagons as their bases. The smaller sandbox has a base area of 1,146 in² and is filled 10 inches deep with sand.

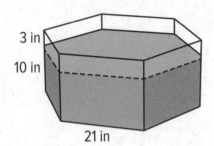

1. It took 14 bags of sand to fill the small sandbox to this depth. What volume of sand comes in one bag? (Round to the nearest whole cubic inch.)

2. The daycare manager wants to add 3 more inches to the depth of the sand in the small sandbox. How many bags of sand will they need to buy?

3. The daycare manager also wants to add 3 more inches to the depth of the sand in the large sandbox. The base of the large sandbox is a scaled copy of the base of the small sandbox, with a scale factor of 1.5. How many bags of sand will they need to buy for the large sandbox?

4. A lawn and garden store is selling 6 bags of sand for $19.50. How much will they spend to buy all the new sand for both sandboxes?

NAME _____ DATE _____ PERIOD _____

Summary
Applying Volume and Surface Area

Suppose we wanted to make a concrete bench like the one shown. If we know that the finished bench has a volume of 10 ft^3 and a surface area of 44 ft^2 we can use this information to solve problems about the bench.

For example,

* How much does the bench weigh?

* How long does it take to wipe the whole bench clean?

* How much will the materials cost to build the bench and to paint it?

To figure out how much the bench weighs, we can use its volume, 10 ft^3. Concrete weighs about 150 pounds per cubic foot, so this bench weighs about 1,500 pounds, because 10 • 150 = 1,500.

To figure out how long it takes to wipe the bench clean, we can use its surface area, 44 ft^2. If it takes a person about 2 seconds per square foot to wipe a surface clean, then it would take about 88 seconds to clean this bench, because 44 • 2 = 88. It may take a little less than 88 seconds, since the surfaces where the bench is touching the ground do not need to be wiped.

Would you use the volume or the surface area of the bench to calculate the cost of the concrete needed to build this bench? And for the cost of the paint?

1. A landscape architect is designing a pool that has this top view:

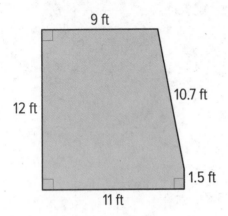

9 ft

10.7 ft

12 ft

1.5 ft

11 ft

a. How much water will be needed to fill this pool 4 feet deep?

b. Before filling up the pool, it gets lined with a plastic liner. How much liner is needed for this pool?

c. Here are the prices for different amounts of plastic liner. How much will all the plastic liner for the pool cost?

Plastic Liner (ft²)	Cost ($)
25	3.75
50	7.50
75	11.25

NAME _____ DATE _____ PERIOD _____

2. Shade in a base of the trapezoidal prism. (The base is not the same as the bottom.) (Lesson 7-13)

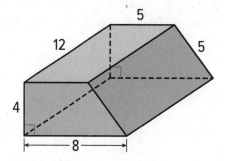

   a. Find the area of the base you shaded.

   b. Find the volume of this trapezoidal prism.

3. For each diagram, decide if y is an increase or a decrease of x. Then determine the percentage that x increased or decreased to result in y. (Lesson 4-9)

A x
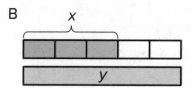

B x

C x
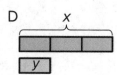

D x

4. Noah is visiting his aunt in Texas. He wants to buy a belt buckle whose price is $25. He knows that the sales tax in Texas is 6.25%. (Lesson 4-10)

a. How much will the tax be on the belt buckle?

b. How much will Noah spend for the belt buckle including the tax?

c. Write an equation that represents the total cost, c, of an item whose price is p.

Lesson 7-17

Building Prisms

NAME _____ DATE _____ PERIOD _____

Learning Goal Let's build a triangular prism from scratch.

Warm Up
17.1 Nets

Here are some nets for various prisms.

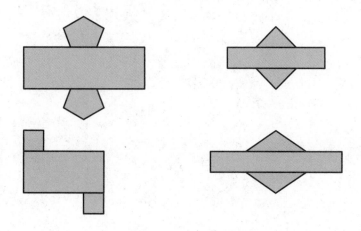

1. What would each net look like when folded?

2. What do you notice about the nets?

Activity

17.2 Making the Base

The base of a triangular prism has one side that is 7 cm long, one side that is 5.5 cm long, and one angle that measures 45°.

1. Draw as many different triangles as you can with these given measurements.

2. Select one of the triangles you have drawn. Measure and calculate to approximate its area. Explain or show your reasoning.

NAME _____ DATE _____ PERIOD _____

Activity
17.3 Making the Prism

Your teacher will give you an incomplete net. Follow these instructions to complete the net and assemble the triangular prism:

1. Draw an identical copy of the triangle you selected in the previous activity along the top of the rectangle, with one vertex on point *A*.

2. Draw another copy of your triangle, flipped upside down, along the bottom of the rectangle, with one vertex on point *C*.

3. Determine how long the rectangle needs to be to wrap all the way around your triangular bases. Pause here so your teacher can review your work.

4. Cut out and assemble your net.

After you finish assembling your triangular prism, answer these questions. Explain or show your reasoning.

1. What is the volume of your prism?

2. What is the surface area of your prism?

3. Stand your prism up so it is sitting on its triangular base.

   a. If you were to cut your prism in half horizontally, what shape would the cross section be?

   b. If you were to cut your prism in half vertically, what shape would the cross section be?

Activity
17.4 Combining Prisms

1. Compare your prism with your partner's prism. What is the same? What is different?

2. Find a way you can put your prism and your partner's prism together to make one new, larger prism. Describe your new prism.

3. Draw the base of your new prism and label the lengths of the sides.

4. As you answer these questions about your new prism, look for ways you can use your calculations from the previous activity to help you. Explain or show your reasoning.

   a. What is the area of its base?

   b. What is its height?

   c. What is its volume?

   d. What is its surface area?

Are you ready for more?

How many identical copies of your prism would it take you to put together a new larger prism in which every dimension was twice as long?

Learning Targets

Lesson	Learning Target(s)
7-1 Relationships of Angles	• I can find unknown angle measures by reasoning about adjacent angles with known measures. • I can recognize when an angle measures 90°, 180°, or 360°.
7-2 Adjacent Angles	• I can find unknown angle measures by reasoning about complementary or supplementary angles. • I can recognize when adjacent angles are complementary or supplementary.
7-3 Nonadjacent Angles	• I can determine if angles that are not adjacent are complementary or supplementary. • I can explain what vertical angles are in my own words.

(continued on the next page)

(continued from the previous page)

Lesson	Learning Target(s)
7-4 Solving for Unknown Angles	• I can reason through multiple steps to find unknown angle measures. • I can recognize when an equation represents a relationship between angle measures.
7-5 Using Equations to Solve for Unknown Angles	• I can write an equation to represent a relationship between angle measures and solve the equation to find unknown angle measures.
7-6 Building Polygons (Part 1)	• I can show that the 3 side lengths that form a triangle cannot be rearranged to form a different triangle. • I can show that the 4 side lengths that form a quadrilateral can be rearranged to form different quadrilaterals.
7-7 Building Polygons (Part 2)	• I can reason about a figure with an unknown angle. • I can show whether or not 3 side lengths will make a triangle.

Lesson	Learning Target(s)
7-8 Triangles with 3 Common Measures	• I understand that changing which sides and angles are next to each other can make different triangles.
7-9 Drawing Triangles (Part 1)	• Given two angle measures and one side length, I can draw different triangles with these measurements or show that these measurements determine one unique triangle or no triangle.
7-10 Drawing Triangles (Part 2)	• Given two side lengths and one angle measure, I can draw different triangles with these measurements or show that these measurements determine one unique triangle or no triangle.
7-11 Slicing Solids	• I can explain that when a three dimensional figure is sliced it creates a face that is two dimensional. • I can picture different cross sections of prisms and pyramids.

(continued on the next page)

Lesson	Learning Target(s)
7-12 Volume of Right Prisms	• I can explain why the volume of a prism can be found by multiplying the area of the base and the height of the prism.
7-13 Decomposing Bases for Area	• I can calculate the volume of a prism with a complicated base by decomposing the base into quadrilaterals or triangles.
7-14 Surface Area of Right Prisms	• I can find and use shortcuts when calculating the surface area of a prism. • I can picture the net of a prism to help me calculate its surface area.
7-15 Distinguishing Volume and Surface Area	• I can decide whether I need to find the surface area or volume when solving a problem about a real-world situation.

Lesson	Learning Target(s)
7-16 Applying Volume and Surface Area	• I can solve problems involving the volume and surface area of children's play structures.
7-17 Building Prisms	• I can build a triangular prism from scratch.

Notes:

_____

_____

_____

_____

_____

_____

_____

_____

_____

_____

_____

(continued on the next page)

(continued from the previous page)

Unit 8

Probability and Sampling

View Stock/View Stock RF/Getty Images

Go team! In this unit, you'll explore how you can use sampling to compare the heights of athletes on gymnastics and volleyball teams.

Topics

- Probabilities of Single-Step Events
- Probabilities of Multi-Step Events
- Sampling
- Using Samples
- Let's Put It to Work

Unit 8

Probability and Sampling

Lesson 8-1

Mystery Bags

NAME _____ DATE _____ PERIOD _____

Learning Goal Let's make predictions based on what we know.

 ## Warm Up
1.1 Going Fishing

Andre and his dad have been fishing for 2 hours. In that time, they have caught 9 bluegills and 1 yellow perch.

The next time Andre gets a bite, what kind of fish do you think it will be? Explain your reasoning.

 ## Activity
1.2 Playing the Block Game

Your teacher will give your group a bag of colored blocks.

1. Follow these instructions to play one round of the game.

   a. Everyone in the group records the color written on the bag in the first column of the table.

   b. Without looking in the bag, one person takes out one of the blocks and shows it to the group.

   c. If they get a block that is the same color as the bag, they earn:

   • 1 point during round 1

   • 2 points during round 2

   • 3 points during round 3

   d. Next, they put the block back into the bag, shake the bag to mix up the blocks, and pass the bag to the next person in the group.

   e. Repeat these steps until everyone in your group has had 4 turns.

2. At the end of the round, record each person's score in the table.

	What color bag?	Person 1's Score	Person 2's Score	Person 3's Score	Person 4's Score
Round 1					
Round 2					
Round 3					

3. Pause here so your teacher can give you a new bag of blocks for the next round.

4. Repeat the previous steps to play rounds 2 and 3 of the game.

5. After you finish playing all 3 rounds, calculate the total score for each person in your group.

Are you ready for more?

Tyler's class played the block game using purple, orange, and yellow bags of blocks.

- During round 1, Tyler's group picked 4 purple blocks and 12 blocks of other colors.

- During round 2, Tyler's group picked 11 orange blocks and 5 blocks of other colors.

- During round 3, Tyler forgot to record how many yellow blocks his group picked.

For a final round, Tyler's group can pick one block from any of the three bags. Tyler's group decides that picking from the orange bag would give them the best chance of winning, and that picking from the purple bag would give them the worst chance of winning. What results from the yellow bag could have lead Tyler's group to this conclusion? Explain your reasoning.

NAME _____ DATE _____ PERIOD _____

Summary
Mystery Bags

One of the main ways that humans learn is by repeating experiments and observing the results.

- Babies learn that dropping their cup makes it hit the floor with a loud noise by repeating this action over and over.

- Scientists learn about nature by observing the results of repeated experiments again and again.

With enough data about the results of experiments, we can begin to predict what may happen if the experiment is repeated in the future.

For example, a baseball player who has gotten a hit 33 out of 100 times at bat might be expected to get a hit about 33% of his times at bat in the future as well.

In some cases, we can predict the chances of things happening based on our knowledge of the situation.

For example, a coin should land heads up about 50% of the time due to the symmetry of the coin.

In other cases, there are too many unknowns to predict the chances of things happening.

For example, the chances of rain tomorrow are based on similar weather conditions we have observed in the past. In these situations, we can experiment, using past results to estimate chances.

Practice
Mystery Bags

1. Lin is interested in how many of her classmates watch her favorite TV show, so she starts asking around at lunch. She gets the following responses:

yes   yes   yes   no   no   no   no   no

no   no   yes   no   no   no

If she asks one more person randomly in the cafeteria, do you think they will say "yes" or "no"? Explain your reasoning.

2. An engineer tests the strength of a new material by seeing how much weight it can hold before breaking. Previous tests have held these weights in pounds:

1,200   1,400   1,300   1,500   950   1,600   1,100

Do you think that this material will be able to hold more than 1,000 pounds in the next test? Explain your reasoning.

NAME _____ DATE _____ PERIOD _____

3. A company tests two new products to make sure they last for more than a year.

- Product 1 had 950 out of 1,000 test items last for more than a year.

- Product 2 had 150 out of 200 last for more than a year.

If you had to choose one of these two products to use for more than a year, which one is more likely to last? Explain your reasoning.

4. Put these numbers in order from least to greatest.

$\frac{1}{2}$ $\qquad$ $\frac{1}{3}$ $\qquad$ $\frac{2}{5}$ $\qquad$ 0.6 $\qquad$ 0.3

5. A small staircase is made so that the horizontal piece of each step is 10 inches long and 25 inches wide. Each step is 5 inches above the previous one. What is the surface area of this staircase? (Lesson 7-15)

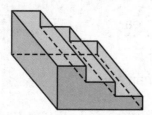

Lesson 8-2

Chance Experiments

NAME _____ DATE _____ PERIOD _____

Learning Goal Let's investigate chance.

Warm Up
2.1 Which Is More Likely?

Which is more likely to happen?

A. When reaching into a dark closet and pulling out one shoe from a pile of 20 pairs of shoes, you pull out a left shoe.

B. When listening to a playlist—which has 5 songs on it—in shuffle mode, the first song on the playlist plays first.

Activity
2.2 How Likely Is It?

1. Label each event with one of these options:

   impossible, unlikely, equally likely as not, likely, certain

   a. You will win the grand prize in a raffle if you purchased 2 out of the 100 tickets.

   b. You will wait less than 10 minutes before ordering at a fast food restaurant.

   c. You will get an even number when you roll a standard number cube.

   d. A four-year-old child is over 6 feet tall.

   e. No one in your class will be late to class next week.

f. The next baby born at a hospital will be a boy.

g. It will snow at our school on July 1.

h. The Sun will set today before 11:00 p.m.

i. Spinning this spinner will result in green.

j. Spinning this spinner will result in red.

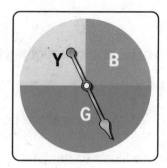

2. Discuss your answers to the previous question with your partner. If you disagree, work to reach an agreement.

3. Invent another situation for each label, for a total of 5 more events.

 Activity

2.3 Take a Chance

Your teacher will have 2 students play a short game.

1. When the first person chose 3 numbers, did they usually win?

2. When the person chose 4 numbers, did you expect them to win? Explain your reasoning.

NAME _____ DATE _____ PERIOD _____

Are you ready for more?

On a game show, there are 3 closed doors. One door has a prize behind it. The contestant chooses one of the doors. The host of the game show, who knows where the prize is located, opens one of the *other* doors which does not have the prize. The contestant can choose to stay with their first choice or switch to the remaining closed door.

1. Do you think it matters if the contestant switches doors or stays?

2. Practice playing the game with your partner and record your results. Whoever is the host starts each round by secretly deciding which door has the prize.

   a. Play 20 rounds where the contestant always stays with their first choice.

   b. Play 20 more rounds where the contestant always switches doors.

3. Did the results from playing the game change your answer to the first question? Explain.

Activity
2.4 Card Sort: Likelihood

1. Your teacher will give you some cards that describe events. Order the events from least likely to most likely.

2. After ordering the first set of cards, pause here so your teacher can review your work. Then, your teacher will give you a second set of cards.

3. Add the new set of cards to the first set so that all of the cards are ordered from least likely to most likely.

Summary
Chance Experiments

A **chance experiment** is something that happens where the outcome is unknown. For example, if we flip a coin, we don't know if the result will be a head or a tail.

An **outcome** of a chance experiment is something that can happen when you do a chance experiment. For example, when you flip a coin, one possible outcome is that you will get a head.

An **event** is a set of one or more outcomes.

We can describe events using these phrases:

- Impossible
- Likely
- Unlikely
- Certain
- Equally likely as not

For example, if you flip a coin:

- It is *impossible* that the coin will turn into a bottle of ketchup.

- It is *unlikely* the coin will land on its edge.

- It is *equally likely as not* that you will get a tail.

- It is *likely* that you will get a head or a tail.

- It is *certain* that the coin will land somewhere.

The *probability* of an event is a measure of the likelihood that an event will occur. We will learn more about probabilities in the lessons to come.

Glossary
_____
chance experiment
event
outcome

NAME _____ DATE _____ PERIOD _____

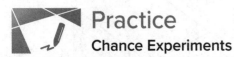

Practice
Chance Experiments

1. The likelihood that Han makes a free throw in basketball is 60%. The likelihood that he makes a 3-point shot is 0.345. Which event is more likely, Han making a free throw or making a 3-point shot? Explain your reasoning.

2. Different events have the following likelihoods. Sort them from least to greatest:

   60%   8 out of 10   0.37   20% $\frac{5}{6}$

3. There are 25 prime numbers between 1 and 100. There are 46 prime numbers between 1 and 200. Which situation is more likely? Explain your reasoning.

   • A computer produces a random number between 1 and 100 that is prime.

   • A computer produces a random number between 1 and 200 that is prime.

4. It takes $4\frac{3}{8}$ cups of cheese, $\frac{7}{8}$ cup of olives, and $2\frac{5}{8}$ cups of sausage to make a signature pizza. How much of each ingredient is needed to make 10 pizzas? Explain or show your reasoning. **(Lesson 4-2)**

5. Here is a diagram of a birdhouse Elena is planning to build. (It is a simplified diagram, since in reality, the sides will have a thickness.) About how many square inches of wood does she need to build this birdhouse? **(Lesson 7-16)**

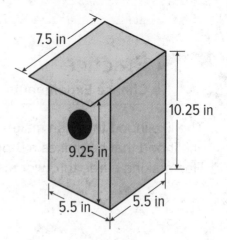

7.5 in

10.25 in

9.25 in

5.5 in   5.5 in

6. Select **all** the situations where knowing the surface area of an object would be more useful than knowing its volume. **(Lesson 7-15)**

(A.) placing an order for tiles to replace the roof of a house

(B.) estimating how long it will take to clean the windows of a greenhouse

(C.) deciding whether leftover soup will fit in a container

(D.) estimating how long it will take to fill a swimming pool with a garden hose

(E.) calculating how much paper is needed to manufacture candy bar wrappers

(F.) buying fabric to sew a couch cover

(G.) deciding whether one muffin pan is enough to bake a muffin recipe

Lesson 8-3

What are Probabilities?

NAME _____ DATE _____ PERIOD _____

Learning Goal Let's find out what's possible.

Warm Up
3.1 Which Game Would You Choose?

Which game would you choose to play? Explain your reasoning.

Game 1: You flip a coin and win if it lands showing heads.

Game 2: You roll a standard number cube and win if it lands showing a number that is divisible by 3.

Activity
3.2 What's Possible?

1. For each situation, list the **sample space** and tell how many outcomes there are.

   a. Han rolls a standard number cube once.

   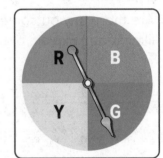

   b. Clare spins this spinner once.

   c. Kiran selects a letter at **random** from the word "MATH."

   d. Mai selects a letter at random from the alphabet.

   e. Noah picks a card at random from a stack that has cards numbered 5 through 20.

2. Next, compare the likelihood of these outcomes. Be prepared to explain your reasoning.

a. Is Clare more likely to have the spinner stop on the red or blue section?

b. Is Kiran or Mai more likely to get the letter T?

c. Is Han or Noah more likely to get a number that is greater than 5?

3. Suppose you have a spinner that is evenly divided showing all the days of the week. You also have a bag of papers that list the months of the year. Are you more likely to spin the current day of the week or pull out the paper with the current month?

Are you ready for more?

Are there any outcomes for two people in this activity that have the same likelihood? Explain or show your reasoning.

NAME _____ DATE _____ PERIOD _____

Activity
3.3 What's in the Bag?

Your teacher will give your group a bag of paper slips with something printed on them. Repeat these steps until everyone in your group has had a turn.

- As a group, guess what is printed on the papers in the bag and record your guess in the table.

- Without looking in the bag, one person takes out one of the papers and shows it to the group.

- Everyone in the group records what is printed on the paper.

- The person who took out the paper puts it back into the bag, shakes the bag to mix up the papers, and passes the bag to the next person in the group.

	Guess the sample space.	What is printed on the paper?
Person 1		
Person 2		
Person 3		
Person 4		

1. How was guessing the sample space the fourth time different from the first?

2. What could you do to get a better guess of the sample space?

3. Look at all the papers in the bag. Were any of your guesses correct?

4. Are all of the possible outcomes equally likely? Explain.

5. Use the sample space to determine the **probability** that a fifth person would get the same outcome as person 1.

Summary

What are Probabilities?

The **probability** of an event is a measure of the likelihood that the event will occur. Probabilities are expressed using numbers from 0 to 1.

If the probability is 0, that means the event is impossible. For example, when you flip a coin, the probability that it will turn into a bottle of ketchup is 0. The closer the probability of some event is to 0, the less likely it is.

If the probability is 1, that means the event is certain. For example, when you flip a coin, the probability that it will land somewhere is 1. The closer the probability of some event is to 1, the more likely it is.

If we list all of the possible outcomes for a chance experiment, we get the **sample space** for that experiment. For example, the sample space for rolling a standard number cube includes six outcomes: 1, 2, 3, 4, 5, and 6. The probability that the number cube will land showing the number 4 is $\frac{1}{6}$.

In general, if all outcomes in an experiment are equally likely and there are n possible outcomes, then the probability of a single outcome is $\frac{1}{n}$.

Sometimes we have a set of possible outcomes and we want one of them to be selected at **random**. That means that we want to select an outcome in a way that each of the outcomes is *equally likely*. For example, if two people both want to read the same book, we could flip a coin to see who gets to read the book first.

Glossary

probability
random
sample space

NAME _____ DATE _____ PERIOD _____

Practice
What are Probabilities?

1. List the *sample space* for each chance experiment.

   a. Flipping a coin

   b. Selecting a random season of the year

   c. Selecting a random day of the week

2. A computer randomly selects a letter from the alphabet.

   a. How many different outcomes are in the sample space?

   b. What is the probability the computer produces the first letter of your first name?

3. What is the probability of selecting a random month of the year and getting a month that starts with the letter "J?" If you get stuck, consider listing the sample space.

4. *E* represents an object's weight on Earth and *M* represents that same object's weight on the Moon. The equation $M = \frac{1}{6}E$ represents the relationship between these quantities. **(Lesson 2-4)**

   a. What does the $\frac{1}{6}$ represent in this situation?

   b. Give an example of what a person might weigh on Earth and on the Moon.

5. Here is a diagram of the base of a bird feeder which is in the shape of a pentagonal prism. Each small square on the grid is 1 square inch.

   The distance between the two bases is 8 inches. What will be the volume of the completed bird feeder?

   (Lesson 7-13)

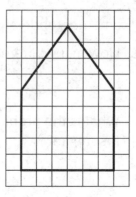

6. Find the surface area of the triangular prism. **(Lesson 7-14)**

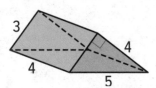

Lesson 8-4

Estimating Probabilities through Repeated Experiments

NAME _____ DATE _____ PERIOD _____

Learning Goal Let's do some experimenting.

Warm Up
4.1 Decimals on the Number Line

1. Locate and label these numbers on the number line.

   a. 0.5   b. 0.75   c. 0.33

   d. 0.67   e. 0.25

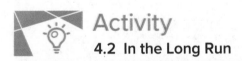

2. Choose one of the numbers from the previous question. Describe a game in which that number represents your probability of winning.

Activity
4.2 In the Long Run

Mai plays a game in which she only wins if she rolls a 1 or a 2 with a standard number cube.

1. List the outcomes in the sample space for rolling the number cube.

2. What is the probability Mai will win the game? Explain your reasoning.

3. If Mai is given the option to flip a coin and win if it comes up heads, is that a better option for her to win?

4. With your group, follow these instructions 10 times to create the graph.

- One person rolls the number cube. Everyone records the outcome.

- Calculate the fraction of rolls that are a win for Mai so far. Approximate the fraction with a decimal value rounded to the hundredths place. Record both the fraction and the decimal in the last column of the table.

- On the graph, plot the number of rolls and the fraction that were wins.

- Pass the number cube to the next person in the group.

Roll	Outcome	Total Number of Wins for Mai	Fraction of Games Played that are Wins
1			
2			
3			
4			
5			
6			
7			
8			
9			
10			

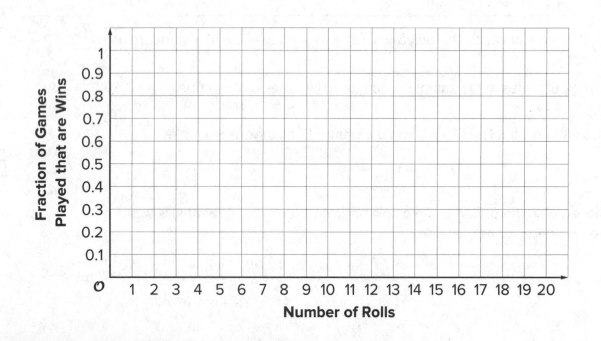

NAME _____ DATE _____ PERIOD _____

5. What appears to be happening with the points on the graph?

6. Respond to the following questions.

    a. After 10 rolls, what fraction of the total rolls were a win?

    b. How close is this fraction to the probability that Mai will win?

7. Roll the number cube 10 more times. Records your results in this table and on the graph from earlier.

Roll	Outcome	Total Number of Wins for Mai	Fraction of Games Played that are Wins
11			
12			
13			
14			
15			
16			
17			
18			
19			
20			

8. Respond to the following questions.

    a. After 20 rolls, what fraction of the total rolls were a win?

    b. How close is this fraction to the probability that Mai will win?

Activity

4.3 Due For a Win

1. For each situation, do you think the result is surprising or not? Is it possible? Be prepared to explain your reasoning.

   a. You flip the coin once, and it lands heads up.

   b. You flip the coin twice, and it lands heads up both times.

   c. You flip the coin 100 times, and it lands heads up all 100 times.

2. If you flip the coin 100 times, how many times would you expect the coin to land heads up? Explain your reasoning.

3. If you flip the coin 100 times, what are some other results that would not be surprising?

4. You've flipped the coin 3 times, and it has come up heads once. The cumulative fraction of heads is currently $\frac{1}{3}$. If you flip the coin one more time, will it land heads up to make the cumulative fraction $\frac{2}{4}$?

NAME _____ DATE _____ PERIOD _____

Summary
Estimating Probabilities through Repeated Experiments

A probability for an event represents the proportion of the time we expect that event to occur in the long run.

For example, the probability of a coin landing heads up after a flip is $\frac{1}{2}$, which means that if we flip a coin many times, we expect that it will land heads up about half of the time.

Even though the probability tells us what we should expect if we flip a coin many times, that doesn't mean we are more likely to get heads if we just got three tails in a row.

The chances of getting heads are the same every time we flip the coin, no matter what the outcome was for past flips.

1. A carnival game has 160 rubber ducks floating in a pool. The person playing the game takes out one duck and looks at it.

   • If there's a red mark on the bottom of the duck, the person wins a small prize.

   • If there's a blue mark on the bottom of the duck, the person wins a large prize.

   • Many ducks do not have a mark.

   After 50 people have played the game, only 3 of them have won a small prize, and none of them have won a large prize.

   Estimate the number of the 160 ducks that you think have red marks on the bottom. Then estimate the number of ducks you think have blue marks. Explain your reasoning.

2. Lin wants to know if flipping a quarter really does have a probability of $\frac{1}{2}$ of landing heads up, so she flips a quarter 10 times. It lands heads up 3 times and tails up 7 times. Has she proven that the probability is not $\frac{1}{2}$? Explain your reasoning.

NAME _____ DATE _____ PERIOD _____

3. A spinner has four equal sections, with one letter from the word "MATH" in each section.

    a. You spin the spinner 20 times. About how many times do you expect it will land on A?

    b. You spin the spinner 80 times. About how many times do you expect it will land on something other than A? Explain your reasoning.

4. A spinner is spun 40 times for a game. Here is a graph showing the fraction of games that are wins under some conditions.

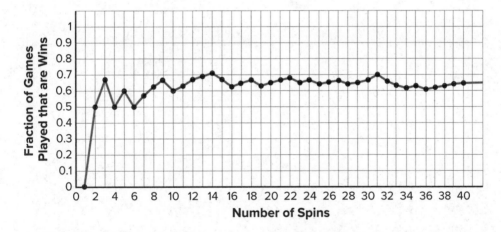

Estimate the probability of a spin winning this game based on the graph.

5. Which event is more likely: rolling a standard number cube and getting an even number, or flipping a coin and having it land heads up? (Lesson 8-2)

6. Noah will select a letter at random from the word "FLUTE." Lin will select a letter at random from the word "CLARINET."

Which person is more likely to pick the letter "E?" Explain your reasoning. (Lesson 8-3)

Lesson 8-5

More Estimating Probabilities

NAME _____ DATE _____ PERIOD _____

Learning Goal Let's estimate some probabilities.

Warm Up
5.1 Is it Likely?

1. If the weather forecast calls for a 20% chance of light rain tomorrow, would you say that it is likely to rain tomorrow?

2. If the probability of a tornado today is $\frac{1}{10}$, would you say that there will likely be a tornado today?

3. If the probability of snow this week is 0.85, would you say that it is likely to snow this week?

Activity
5.2 Making My Head Spin

Your teacher will give you 4 spinners. Make sure each person in your group uses a different spinner.

1. Spin your spinner 10 times, and record your outcomes.

2. Did you get all of the different possible outcomes in your 10 spins?

3. What fraction of your 10 spins landed on 3?

4. Next, share your outcomes with your group, and record their outcomes.

   a. Outcomes for spinner A:

   b. Outcomes for spinner B:

   c. Outcomes for spinner C:

   d. Outcomes for spinner D:

5. Do any of the spinners have the same sample space? If so, do they have the same probabilities for each number to result?

6. For each spinner, what is the probability that it lands on the number 3? Explain or show your reasoning.

7. For each spinner, what is the probability that it lands on something other than the number 3? Explain or show your reasoning.

8. Noah put spinner D on top of his closed binder and spun it 10 times. It never landed on the number 1. How might you explain why this happened?

9. Han put spinner C on the floor and spun it 10 times. It never landed on the number 3, so he says that the probability of getting a 3 is 0. How might you explain why this happened?

NAME _____ DATE _____ PERIOD _____

Are you ready for more?

Design a spinner that has a $\frac{2}{3}$ probability of landing on the number 3. Explain how you could precisely draw this spinner.

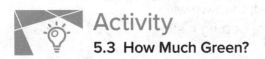

Activity
5.3 How Much Green?

Your teacher will give you a bag of blocks that are different colors. Do not look into the bag or take out more than 1 block at a time. Repeat these steps until everyone in your group has had 4 turns.

- Take one block out of the bag and record whether or not it is green.

- Put the block back into the bag, and shake the bag to mix up the blocks.

- Pass the bag to the next person in the group.

1. What do you think is the probability of taking out a green block from this bag? Explain or show your reasoning.

2. How could you get a better estimate without opening the bag?

Suppose a bag contains 5 blocks. If we select a block at random from the bag, then the probability of getting any one of the blocks is $\frac{1}{5}$.

Now suppose a bag contains 5 blocks. Some of the blocks have a star, and some have a moon. If we select a block from the bag, then we will either get a star block or a moon block. The probability of getting a star block depends on how many there are in the bag.

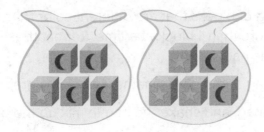

In this example, the probability of selecting a star block at random from the first bag is $\frac{1}{5}$, because it contains only 1 star block. (The probability of getting a moon block is $\frac{4}{5}$.) The probability of selecting a star block at random from the second bag is $\frac{3}{5}$, because it contains 3 star blocks. (The probability of getting a moon block from this bag is $\frac{2}{5}$.)

This shows that two experiments can have the same sample space, but different probabilities for each outcome.

NAME _____ DATE _____ PERIOD _____

 ## Practice
More Estimating Probabilities

1. What is the same about these two experiments? What is different?
   - Selecting a letter at random from the word "ALABAMA"
   - Selecting a letter at random from the word "LAMB"

2. Andre picks a block out of a bag 60 times and notes that 43 of them were green.

   a. What should Andre estimate for the probability of picking out a green block from this bag?

   b. Mai looks in the bag and sees that there are 6 blocks in the bag. Should Andre change his estimate based on this information? If so, what should the new estimate be? If not, explain your reasoning.

3. Han has a number cube that he suspects is not so standard.
   - Han rolls the cube 100 times, and it lands on a six 40 times.
   - Kiran rolls the cube 50 times, and it lands on a six 21 times.
   - Lin rolls the cube 30 times, and it lands on a six 11 times.

   Based on these results, is there evidence to help prove that this cube is not a standard number cube? Explain your reasoning.

4. A textbook has 428 pages numbered in order starting with 1. You flip to a random page in the book in a way that it is equally likely to stop at any of the pages. (Lesson 8-3)

   a. What is the sample space for this experiment?

   b. What is the probability that you turn to page 45?

   c. What is the probability that you turn to an even numbered page?

   d. If you repeat this experiment 50 times, about how many times do you expect you will turn to an even numbered page?

5. A rectangular prism is cut along a diagonal on each face to create two triangular prisms. The distance between A and B is 5 inches.

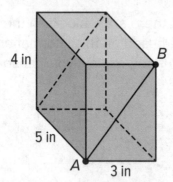

What is the surface area of the original rectangular prism? What is the total surface area of the two triangular prisms together? (Lesson 7-15)

Lesson 8-6

Estimating Probabilities using Simulation

NAME _____ DATE _____ PERIOD _____

Learning Goal Let's simulate real-world situations.

 Warm Up

6.1 Which One Doesn't Belong: Spinners

Which spinner doesn't belong?

Spinner A   **Spinner C**

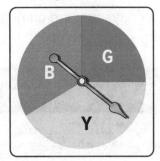

Spinner B   **Spinner D**

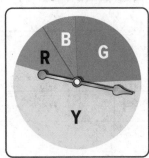
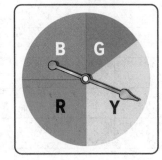

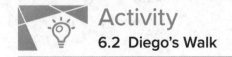

Activity

6.2 Diego's Walk

Your teacher will give your group the supplies for one of the three different simulations. Follow these instructions to simulate 15 days of Diego's walk. The first 3 days have been done for you.

- Simulate one day:

  - If your group gets a bag of papers, reach into the bag, and select one paper without looking inside.

  - If your group gets a spinner, spin the spinner, and see where it stops.

  - If your group gets two number cubes, roll both cubes, and add the numbers that land face up. A sum of 2–8 means Diego has to wait.

- Record in the table whether or not Diego had to wait more than 1 minute.

- Calculate the total number of days and the cumulative fraction of days that Diego has had to wait so far.

Day	Does Diego have to wait more than 1 minute?	Total Number of Days Diego Had to Wait	Fraction of Days Diego Had to Wait
1	no	0	$\frac{0}{1} = 0.00$
2	yes	1	$\frac{1}{2} = 0.50$
3	yes	2	$\frac{2}{3} \approx 0.67$
4			
5			
6			
7			
8			
9			
10			
11			
12			
13			
14			
15			

NAME _____ DATE _____ PERIOD _____

- On the graph, plot the number of days and the fraction that Diego has had to wait. Connect each point by a line.

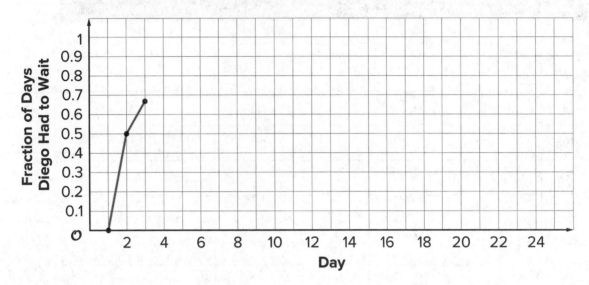

- If your group has the bag of papers, put the paper back into the bag, and shake the bag to mix up the papers.

- Pass the supplies to the next person in the group.

1. Based on the data you have collected, do you think the fraction of days Diego has to wait after the 16th day will be closer to 0.9 or 0.7? Explain or show your reasoning.

2. Continue the simulation for 10 more days. Record your results in this table and on the graph from earlier.

Day	Does Diego have to wait more than 1 minute?	Total Number of Days Diego Had to Wait	Fraction of Days Diego Had to Wait
16			
17			
18			
19			
20			
21			
22			
23			
24			
25			

3. What do you notice about the graph?

4. Based on the graph, estimate the probability that Diego will have to wait more than 1 minute to cross the crosswalk.

NAME _____ DATE _____ PERIOD _____

Are you ready for more?

Let's look at why the values tend to not change much after doing the simulation many times.

1. After doing the simulation 4 times, a group finds that Diego had to wait 3 times. What is an estimate for the probability Diego has to wait based on these results?

   a. If this group does the simulation 1 more time, what are the two possible outcomes for the fifth simulation?

   b. For each possibility, estimate the probability Diego has to wait.

   c. What are the differences between the possible estimates after 5 simulations and the estimate after 4 simulations?

2. After doing the simulation 20 times, this group finds that Diego had to wait 15 times. What is an estimate for the probability Diego has to wait based on these results?

   a. If this group does the simulation 1 more time, what are the two possible outcomes for the twenty-first simulation?

   b. For each possibility, estimate the probability Diego has to wait.

   c. What are the differences between the possible estimates after 21 simulations and the estimate after 20 simulations?

3. Use these results to explain why a single result after many simulations does not affect the estimate as much as a single result after only a few simulations.

Activity

6.3 Designing Experiments

For each situation, describe a chance experiment that would fairly represent it.

1. Six people are going out to lunch together. One of them will be selected at random to choose which restaurant to go to. Who gets to choose?

2. After a robot stands up, it is equally likely to step forward with its left foot or its right foot. Which foot will it use for its first step?

3. In a computer game, there are three tunnels. Each time the level loads, the computer randomly selects one of the tunnels to lead to the castle. Which tunnel is it?

4. Your school is taking 4 buses of students on a field trip. Will you be assigned to the same bus that your math teacher is riding on?

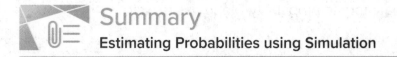

Summary

Estimating Probabilities using Simulation

Sometimes it is easier to estimate a probability by doing a *simulation*.

- A simulation is an experiment that approximates a situation in the real world.

- Simulations are useful when it is hard or time-consuming to gather enough information to estimate the probability of some event.

For example, imagine Andre has to transfer from one bus to another on the way to his music lesson. Most of the time he makes the transfer just fine, but sometimes the first bus is late and he misses the second bus. We could set up a simulation with slips of paper in a bag. Each paper is marked with a time when the first bus arrives at the transfer point. We select slips at random from the bag. After many trials, we calculate the fraction of the times that he missed the bus to estimate the probability that he will miss the bus on a given day.

Glossary

simulation

NAME _____ DATE _____ PERIOD _____

Practice
Estimating Probabilities using Simulation

1. The weather forecast says there is a 75% chance it will rain later today.

   a. Draw a spinner you could use to simulate this probability.

   b. Describe another way you could simulate this probability.

2. An experiment will produce one of ten different outcomes with equal probability for each. Why would using a standard number cube to simulate the experiment be a bad choice?

3. An ice cream shop offers 40 different flavors. To simulate the most commonly chosen flavor, you could write the name of each flavor on a piece of paper and put it in a bag. Draw from the bag 100 times, and see which flavor is chosen the most. This simulation is not a good way to figure out the most-commonly chosen flavor. Explain why.

4. Each set of three numbers represents the lengths, in units, of the sides of a triangle. Which set can *not* be used to make a triangle? **(Lesson 7-7)**

   A. 7, 6, 14

   B. 4, 4, 4

   C. 6, 6, 2

   D. 7, 8, 13

5. There is a proportional relationship between a volume measured in cups and the same volume measured in tablespoons. 48 tablespoons is equivalent to 3 cups, as shown in the graph. (Lesson 2-14)

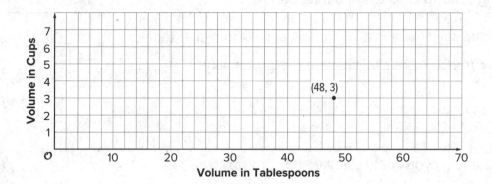

a. Plot and label some more points that represent the relationship.

b. Use a straightedge to draw a line that represents this proportional relationship.

c. For which value y is $(1, y)$ on the line you just drew?

d. What is the constant of proportionality for this relationship?

e. Write an equation representing this relationship. Use c for cups and t for tablespoons.

Lesson 8-7

Simulating Multi-Step Experiments

NAME _____ DATE _____ PERIOD _____

Learning Goal Let's simulate more complicated events.

Warm Up
7.1 Notice and Wonder: Ski Business

What do you notice? What do you wonder?

Activity
7.2 Alpine Zoom

Alpine Zoom is a ski business. To make money over spring break, they need it to snow at least 4 out of the 10 days. The weather forecast says there is a $\frac{1}{3}$ chance it will snow each day during the break.

1. Describe a chance experiment that you could use to simulate whether it will snow on the first day of spring break.

2. How could this chance experiment be used to determine whether Alpine Zoom will make money?

Pause here so your teacher can give you the supplies for a simulation.

3. Simulate the weather for 10 days to see if Alpine Zoom will make money over spring break. Record your results in the first row of the table.

	Day 1	Day 2	Day 3	Day 4	Day 5	Day 6	Day 7	Day 8	Day 9	Day 10	Did they make money over spring break?
Simulation 1											
Simulation 2											
Simulation 3											
Simulation 4											
Simulation 5											

4. Repeat the previous step 4 more times. Record your results in the other rows of the table.

5. Based on your group's simulations, estimate the probability that Alpine Zoom will make money.

NAME _____ DATE _____ PERIOD _____

Activity
7.3 Kiran's Game

Kiran invents a game that uses a board with alternating black and white squares. A playing piece starts on a white square and must advance 4 squares to the other side of the board within 5 turns to win the game.

For each turn, the player draws a block from a bag containing 2 black blocks and 2 white blocks. If the block color matches the color of the next square on the board, the playing piece moves onto it. If it does not match, the playing piece stays on its current square.

1. Take turns playing the game until each person in your group has played the game twice.

2. Use the results from all the games your group played to estimate the probability of winning Kiran's game.

3. Do you think your estimate of the probability of winning is a good estimate? How could it be improved?

Are you ready for more?

How would each of these changes, on its own, affect the probability of winning the game?

1. Change the rules so that the playing piece must move 7 spaces within 8 moves.

2. Change the board so that all the spaces are black.

3. Change the blocks in the bag to 3 black blocks and 1 white block.

Activity

7.4: Simulation Nation

Match each situation to a simulation.

Situations	Simulations

a. In a small lake, 25% of the fish are female. You capture a fish, record whether it is male or female, and toss the fish back into the lake. If you repeat this process 5 times, what is the probability that at least 3 of the 5 fish are female?

i. Toss a standard number cube 2 times and record the outcomes. Repeat this process many times and find the proportion of the simulations in which a 1 or 2 appeared both times to estimate the probability.

b. Elena makes about 80% of her free throws. Based on her past successes with free throws, what is the probability that she will make exactly 4 out of 5 free throws in her next basketball game?

ii. Make a spinner with four equal sections labeled 1, 2, 3, and 4. Spin the spinner 5 times and record the outcomes. Repeat this process many times and find the proportion of the simulations in which a 4 appears 3 or more times to estimate the probability.

c. On a game show, a contestant must pick one of three doors. In the first round, the winning door has a vacation. In the second round, the winning door has a car. What is the probability of winning a vacation and a car?

iii. Toss a fair coin 4 times and record the outcomes. Repeat this process many times, and find the proportion of the simulations in which exactly 3 heads appear to estimate the probability.

d. Your choir is singing in 4 concerts. You and one of your classmates both learned the solo. Before each concert, there is an equal chance the choir director will select you or the other student to sing the solo. What is the probability that you will be selected to sing the solo in exactly 3 of the 4 concerts?

iv. Place 8 blue chips and 2 red chips in a bag. Shake the bag, select a chip, record its color, and then return the chip to the bag. Repeat the process 4 more times to obtain a simulated outcome. Then repeat this process many times and find the proportion of the simulations in which exactly 4 blues are selected to estimate the probability.

NAME _____ DATE _____ PERIOD _____

Summary
Simulating Multi-Step Experiments

The more complex a situation is, the harder it can be to estimate the probability of a particular event happening.

Well-designed simulations are a way to estimate a probability in a complex situation, especially when it would be difficult or impossible to determine the probability from reasoning alone.

To design a good simulation, we need to know something about the situation. For example, if we want to estimate the probability that it will rain every day for the next three days, we could look up the weather forecast for the next three days. Here is a table showing a weather forecast:

	Today (Tuesday)	Wednesday	Thursday	Friday
Probability of Rain	0.2	0.4	0.5	0.9

We can set up a simulation to estimate the probability of rain each day with three bags.

- In the first bag, we put 4 slips of paper that say "rain" and 6 that say "no rain."

- In the second bag, we put 5 slips of paper that say "rain" and 5 that say "no rain."

- In the third bag, we put 9 slips of paper that say "rain" and 1 that says "no rain."

Then we can select one slip of paper from each bag and record whether or not there was rain on all three days.
If we repeat this experiment many times, we can estimate the probability that there will be rain on all three days by dividing the number of times all three slips said "rain" by the total number of times we performed the simulation.

Practice
Simulating Multi-Step Experiments

1. Priya's cat is pregnant with a litter of 5 kittens. Each kitten has a 30% chance of being chocolate brown. Priya wants to know the probability that at least two of the kittens will be chocolate brown.

   To simulate this, Priya put 3 white cubes and 7 green cubes in a bag. For each trial, Priya pulled out and returned a cube 5 times. Priya conducted 12 trials.
   Here is a table with the results.

Trial Number	Outcome
1	ggggg
2	gggwg
3	wgwgw
4	gwggg
5	gggwg
6	wwggg
7	gwggg
8	ggwgw
9	wwwgg
10	ggggw
11	wggwg
12	gggwg

   a. How many successful trials were there? Describe how you determined if a trial was a success.

   b. Based on this simulation, estimate the probability that *exactly* two kittens will be chocolate brown.

   c. Based on this simulation, estimate the probability that *at least* two kittens will be chocolate brown.

   d. Write and answer another question Priya could answer using this simulation.

   e. How could Priya increase the accuracy of the simulation?

NAME _____ DATE _____ PERIOD _____

2. A team has a 75% chance to win each of the 3 games they will play this week. Clare simulates the week of games by putting 4 pieces of paper in a bag, 3 labeled "win" and 1 labeled "lose." She draws a paper, writes down the result, then replaces the paper and repeats the process two more times. Clare gets the result: win, win, lose. What can Clare do to estimate the probability the team will win at least 2 games?

3. Respond to each of the following. (Lesson 8-5)

   a. List the sample space for selecting a letter at random from the word "PINEAPPLE."

   b. A letter is randomly selected from the word "PINEAPPLE." Which is more likely, selecting "E" or selecting "P"? Explain your reasoning.

4. On a graph of side length of a square vs. its perimeter, a few points are plotted. **(Lesson 2-11)**

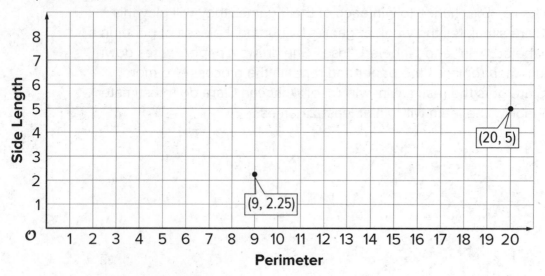

a. Add at least two more ordered pairs to the graph.

b. Is there a proportional relationship between the perimeter and side length? Explain how you know.

Lesson 8-8

Keeping Track of all Possible Outcomes

NAME _____ DATE _____ PERIOD _____

Learning Goal Let's explore sample spaces for experiments with multiple parts.

Warm Up
8.1 How Many Different Meals?

How many different meals are possible if each meal includes one main course, one side dish, and one drink?

Main Courses	Side Dishes	Drinks
grilled chicken	salad	milk
turkey sandwich	applesauce	juice
pasta salad	—	water

Consider the experiment: Flip a coin, and then roll a number cube.

Elena, Kiran, and Priya each use a different method for finding the sample space of this experiment.

- Elena carefully writes a list of all the options: Heads 1, Heads 2, Heads 3, Heads 4, Heads 5, Heads 6, Tails 1, Tails 2, Tails 3, Tails 4, Tails 5, Tails 6.

- Kiran makes a table:

	1	2	3	4	5	6
H	H1	H2	H3	H4	H5	H6
T	T1	T2	T3	T4	T5	T6

- Priya draws a tree with branches in which each pathway represents a different outcome:

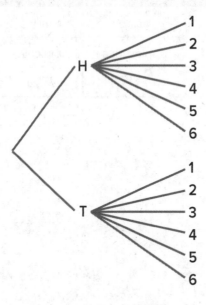

1. Compare the three methods. What is the same about each method? What is different? Be prepared to explain why each method produces all the different outcomes without repeating any.

2. Which method do you prefer for this situation?

Pause here so your teacher can review your work.

NAME _____ DATE _____ PERIOD _____

3. Find the sample space for each of these experiments using any method. Make sure you list every possible outcome without repeating any.

   a. Flip a dime, then flip a nickel, and then flip a penny. Record whether each lands heads or tails up.

   b. Han's closet has: a blue shirt, a gray shirt, a white shirt, blue pants, khaki pants, and black pants. He must select one shirt and one pair of pants to wear for the day.

   c. Spin a color, and then spin a number.

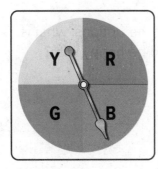 

   d. Spin the hour hand on an analog clock, and then choose a.m. or p.m.

 **Activity**

8.3 How Many Sandwiches?

1. A submarine sandwich shop makes sandwiches with one kind of bread, one protein, one choice of cheese, and *two* vegetables. How many different sandwiches are possible? Explain your reasoning. You do not need to write out the sample space.

   - Breads: Italian, white, wheat

   - Proteins: Tuna, ham, turkey, beans

   - Cheese: Provolone, Swiss, American, none

   - Vegetables: Lettuce, tomatoes, peppers, onions, pickles

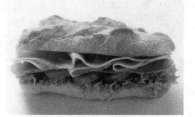

2. Andre knows he wants a sandwich that has ham, lettuce, and tomatoes on it. He doesn't care about the type of bread or cheese. How many of the different sandwiches would make Andre happy?

3. If a sandwich is made by randomly choosing each of the options, what is the probability it will be a sandwich that Andre would be happy with?

Are you ready for more?

Describe a situation that involves three parts and has a total of 24 outcomes in the sample space.

NAME _____ DATE _____ PERIOD _____

Summary
Keeping Track of all Possible Outcomes

Sometimes we need a systematic way to count the number of outcomes that are possible in a given situation.

For example, suppose there are 3 people (A, B, and C) who want to run for the president of a club and 4 different people (1, 2, 3, and 4) who want to run for vice president of the club.

We can use a *tree*, a *table*, or an *ordered list* to count how many different combinations are possible for a president to be paired with a vice president.

With a tree, we can start with a branch for each of the people who want to be president. Then for each possible president, we add a branch for each possible vice president, for a total of 3 • 4 = 12 possible pairs. We can also start by counting vice presidents first and then adding a branch for each possible president, for a total of 4 • 3 = 12 possible pairs.

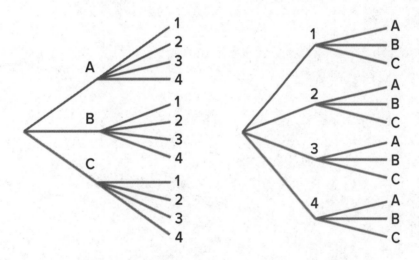

A table can show the same result:

	1	2	3	4
A	A, 1	A, 2	A, 3	A, 4
B	B, 1	B, 2	B, 3	B, 4
C	C, 1	C, 2	C, 3	C, 4

So does this ordered list:

A1, A2, A3, A4, B1, B2, B3, B4, C1, C2, C3, C4

Practice
Keeping Track of all Possible Outcomes

1. Noah is planning his birthday party. Here is a tree showing all of the possible themes, locations, and days of the week that Noah is considering.

    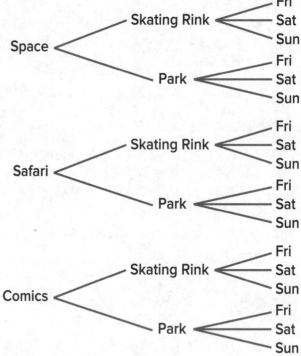

    a. How many themes is Noah considering?

    b. How many locations is Noah considering?

    c. How many days of the week is Noah considering?

    d. One possibility that Noah is considering is a party with a space theme at the skating rink on Sunday. Write two other possible parties Noah is considering.

    e. How many different possible outcomes are in the sample space?

2. For each event, write the sample space and tell how many outcomes there are.

    a. Lin selects one type of lettuce and one dressing to make a salad.

    - Lettuce types: iceberg, romaine
    - Dressings: ranch, Italian, French

NAME _____ DATE _____ PERIOD _____

b. Diego chooses rock, paper, or scissors, and Jada chooses rock, paper, or scissors.

c. Spin these 3 spinners.

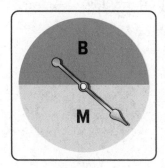 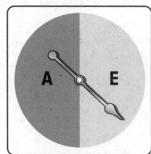 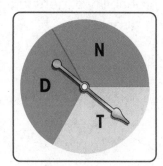

3. A simulation is done to represent kicking 5 field goals in a single game with a 72% probability of making each one. A 1 represents making the kick and a 0 represents missing the kick.

Trial	Result
1	10101
2	11010
3	00011
4	11111
5	10011

Based on these results, estimate the probability that 3 or more kicks are made. **(Lesson 8-7)**

4. There is a bag of 50 marbles.

   • Andre takes out a marble, records its color, and puts it back in. In 4 trials, he gets a green marble 1 time.

   • Jada takes out a marble, records its color, and puts it back in. In 12 trials, she gets a green marble 5 times.

   • Noah takes out a marble, records its color, and puts it back in. In 9 trials, he gets a green marble 3 times.

   Estimate the probability of getting a green marble from this bag. Explain your reasoning. (Lesson 8-4)

Lesson 8-9

Multi-Step Experiments

NAME _____ DATE _____ PERIOD _____

Learning Goal Let's look at probabilities of experiments that have multiple steps.

Warm Up
9.1 True or False?

Is each equation true or false? Explain your reasoning.

1. $8 = (8 + 8 + 8 + 8) \div 3$

2. $(10 + 10 + 10 + 10 + 10) \div 5 = 10$

3. $(6 + 4 + 6 + 4 + 6 + 4) \div 6 = 5$

Activity
9.2 Spinning a Color and Number

The other day, you wrote the sample space for spinning each of these spinners once.

What is the probability of getting:

1. green and 3?

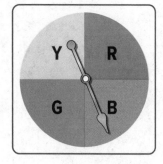

2. blue and any odd number?

3. any color other than red and any number other than 2?

Activity

9.3 Cubes and Coins

The other day you looked at a list, a table, and a tree that showed the sample space for rolling a number cube and flipping a coin.

1. Your teacher will assign you one of these three structures to use to answer these questions. Be prepared to explain your reasoning.

   a. What is the probability of getting tails and a 6?

   b. What is the probability of getting heads and an odd number?

   Pause here so your teacher can review your work.

2. Suppose you roll two number cubes. What is the probability of getting:

   a. both cubes showing the same number?

   b. *exactly* one cube showing an even number?

   c. *at least* one cube showing an even number?

   d. two values that have a sum of 8?

   e. two values that have a sum of 13?

3. Jada flips three quarters. What is the probability that all three will land showing the same side?

NAME _____ DATE _____ PERIOD _____

Activity
9.4 Pick a Card

Imagine there are 5 cards. They are colored red, yellow, green, white, and black. You mix up the cards and select one of them without looking. Then, without putting that card back, you mix up the remaining cards and select another one.

1. Write the sample space and tell how many possible outcomes there are.

2. What structure did you use to write all of the outcomes (list, table, tree, something else)? Explain why you chose that structure.

3. What is the probability that:

   a. you get a white card and a red card (in either order)?

   b. you get a black card (either time)?

   c. you do not get a black card (either time)?

   d. you get a blue card?

   e. you get 2 cards of the same color?

   f. you get 2 cards of different colors?

Are you ready for more?

In a game using five cards numbered 1, 2, 3, 4, and 5, you take two cards and add the values together. If the sum is 8, you win. Would you rather pick a card and put it back before picking the second card, or keep the card in your hand while you pick the second card? Explain your reasoning.

Summary
Multi-Step Experiments

Suppose we have two bags. One contains 1 star block and 4 moon blocks. The other contains 3 star blocks and 1 moon block.

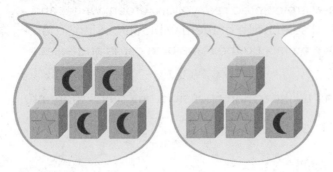

If we select one block at random from each, what is the probability that we will get two star blocks or two moon blocks?

To answer this question, we can draw a tree diagram to see all of the possible outcomes.

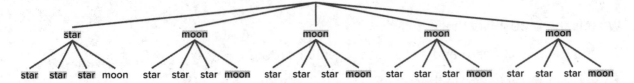

There are 5 · 4 = 20 possible outcomes.
Of these, 3 of them are both stars, and 4 are both moons.
So the probability of getting 2 star blocks or 2 moon blocks is $\frac{7}{20}$.

In general, if all outcomes in an experiment are equally likely, then the probability of an event is the fraction of outcomes in the sample space for which the event occurs.

NAME _____ DATE _____ PERIOD _____

 Practice
Multi-Step Experiments

1. A vending machine has 5 colors (white, red, green, blue, and yellow) of gumballs and an equal chance of dispensing each. A second machine has 4 different animal-shaped rubber bands (lion, elephant, horse, and alligator) and an equal chance of dispensing each. If you buy one item from each machine, what is the probability of getting a yellow gumball and a lion band?

2. The numbers 1 through 10 are put in one bag. The numbers 5 through 14 are put in another bag. When you pick one number from each bag, what is the probability you get the same number?

3. When rolling 3 standard number cubes, the probability of getting all three numbers to match is $\frac{6}{216}$. What is the probability that the three numbers *do not* all match? Explain your reasoning.

4. For each event, write the sample space and tell how many outcomes there are. (Lesson 8-8)

   a. Roll a standard number cube. Then flip a quarter.

   b. Select a month. Then select 2020 or 2025.

5. On a graph of the area of a square vs. its perimeter, a few points are plotted.

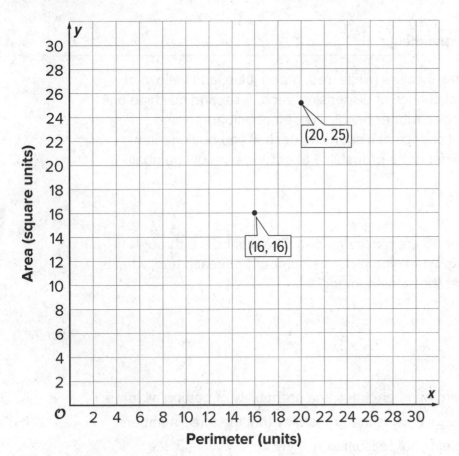

a. Add some more ordered pairs to the graph.

b. Is there a proportional relationship between the area and perimeter of a square? Explain how you know. **(Lesson 2-11)**

Lesson 8-10

Designing Simulations

NAME _____ DATE _____ PERIOD _____

Learning Goal Let's simulate some real-life scenarios.

Warm Up
10.1 Number Talk: Division

Find the value of each expression mentally.

1. $(4.2 + 3) \div 2$

2. $(4.2 + 2.6 + 4) \div 3$

3. $(4.2 + 2.6 + 4 + 3.6) \div 4$

4. $(4.2 + 2.6 + 4 + 3.6 + 3.6) \div 5$

Activity
10.2 Breeding Mice

A scientist is studying the genes that determine the color of a mouse's fur. When two mice with brown fur breed, there is a 25% chance that each baby will have white fur. For the experiment to continue, the scientist needs at least 2 out of 5 baby mice to have white fur.

To simulate this situation, you can flip a coin twice for each baby mouse.

- If the coin lands heads up both times, it represents a mouse with white fur.

- Any other result represents a mouse with brown fur.

1. Simulate 3 litters of 5 baby mice and record your results in the table.

	Mouse 1	Mouse 2	Mouse 3	Mouse 4	Mouse 5	Do at least 2 have white fur?
Simulation 1						
Simulation 2						
Simulation 3						

2. Based on the results from everyone in your group, estimate the probability that the scientist's experiment will be able to continue.

3. How could you improve your estimate?

Are you ready for more?

For a certain pair of mice, the genetics show that each offspring has a probability of $\frac{1}{16}$ that they will be albino. Describe a simulation you could use that would estimate the probability that at least 2 of the 5 offspring are albino.

Activity

10.3 Designing Simulations

Your teacher will give your group a paper describing a situation.

1. Design a simulation that you could use to estimate a probability. Show your thinking. Organize it so it can be followed by others.

2. Explain how you used the simulation to answer the questions posed in the situation.

NAME _____ DATE _____ PERIOD _____

Summary
Designing Simulations

Many real-world situations are difficult to repeat enough times to get an estimate for a probability. If we can find probabilities for parts of the situation, we may be able to simulate the situation using a process that is easier to repeat.

For example, if we know that each egg of a fish in a science experiment has a 13% chance of having a mutation, how many eggs do we need to collect to make sure we have 10 mutated eggs? If getting these eggs is difficult or expensive, it might be helpful to have an idea about how many eggs we need before trying to collect them.

We could simulate this situation by having a computer select random numbers between 1 and 100. If the number is between 1 and 13, it counts as a mutated egg. Any other number would represent a normal egg. This matches the 13% chance of each fish egg having a mutation.

We could continue asking the computer for random numbers until we get 10 numbers that are between 1 and 13. How many times we asked the computer for a random number would give us an estimate of the number of fish eggs we would need to collect.

To improve the estimate, this entire process should be repeated many times. Because computers can perform simulations quickly, we could simulate the situation 1,000 times or more.

Practice
Designing Simulations

1. A rare and delicate plant will only produce flowers from 10% of the seeds planted. To see if it is worth planting 5 seeds to see any flowers, the situation is going to be simulated. Which of these options is the best simulation? For the others, explain why it is not a good simulation.

   - Another plant can be genetically modified to produce flowers 10% of the time. Plant 30 groups of 5 seeds each and wait 6 months for the plants to grow and count the fraction of groups that produce flowers.

   - Roll a standard number cube 5 times. Each time a 6 appears, it represents a plant producing flowers. Repeat this process 30 times and count the fraction of times at least one number 6 appears.

   - Have a computer produce 5 random digits (0 through 9). If a 9 appears in the list of digits, it represents a plant producing flowers. Repeat this process 300 times and count the fraction of times at least one number 9 appears.

   - Create a spinner with 10 equal sections and mark one of them "flowers." Spin the spinner 5 times to represent the 5 seeds. Repeat this process 30 times and count the fraction of times that at least 1 "flower" was spun.

NAME _____ DATE _____ PERIOD _____

2. Jada and Elena learned that 8% of students have asthma. They want to know the probability that in a team of 4 students, at least one of them has asthma. To simulate this, they put 25 slips of paper in a bag. Two of the slips say "asthma." Next, they take four papers out of the bag and record whether at least one of them says "asthma." They repeat this process 15 times.

- Jada says they could improve the accuracy of their simulation by using 100 slips of paper and marking 8 of them.

- Elena says they could improve the accuracy of their simulation by conducting 30 trials instead of 15.

a. Do you agree with either of them? Explain your reasoning.

b. Describe another method of simulating the same scenario.

3. The figure on the left is a trapezoidal prism. The figure on the right represents its base. Find the volume of this prism. **(Lesson 7-13)**

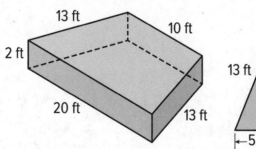
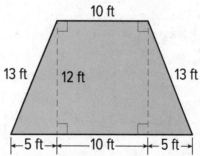

4. Match each expression in the first list with an equivalent expression from the second list. (Lesson 6-22)

Expressions

a. $(8x + 6y) + (2x + 4y)$

b. $(8x + 6y) - (2x + 4y)$

c. $(8x + 6y) - (2x - 4y)$

d. $8x - 6y - 2x + 4y$

e. $8x - 6y + 2x - 4y$

f. $8x - (-6y - 2x + 4y)$

Equivalent Expressions

i. $10(x + y)$

ii. $10(x - y)$

iii. $6(x - \frac{1}{3}y)$

iv. $8x + 6y + 2x - 4y$

v. $8x + 6y - 2x + 4y$

vi. $8x - 2x + 6y - 4y$

Lesson 8-11

Comparing Groups

NAME _____ DATE _____ PERIOD _____

Learning Goal Let's compare two groups.

 Warm Up

11.1 Notice and Wonder: Comparing Heights

What do you notice? What do you wonder?

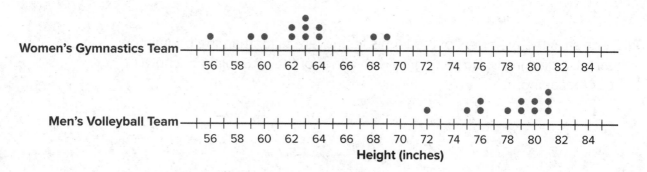

 Activity

11.2 More Team Heights

1. How much taller is the volleyball team than the gymnastics team?

   • Gymnastics team's heights (in inches): 56, 59, 60, 62, 62, 63, 63, 63, 64, 64, 68, 69

   • Volleyball team's heights (in inches): 72, 75, 76, 76, 78, 79, 79, 80, 80, 81, 81, 81

2. Make dot plots to compare the heights of the tennis and badminton teams.

- Tennis team's heights (in inches): 66, 67, 69, 70, 71, 73, 73, 74, 75, 75, 76
- Badminton team's heights (in inches): 62, 62, 65, 66, 68, 71, 73

What do you notice about your dot plots?

3. Elena says the members of the tennis team were taller than the badminton team. Lin disagrees. Do you agree with either of them? Explain or show your reasoning.

Activity
11.3 Family Heights

Compare the heights of these two families. Explain or show your reasoning.

- The heights (in inches) of Noah's family members: 28, 39, 41, 52, 63, 66, 71
- The heights (in inches) of Jada's family members: 49, 60, 68, 70, 71, 73, 77

NAME _____ DATE _____ PERIOD _____

If Jada's family adopts newborn twins who are each 18 inches tall, does this change your thinking? Explain your reasoning.

Activity
11.4 Track Length

Here are three dot plots that represent the lengths, in minutes, of songs on different albums.

A

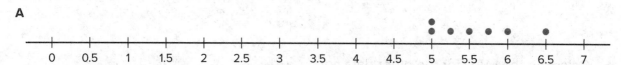

B

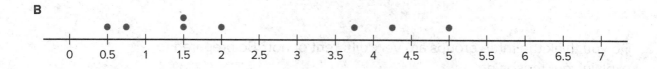

C

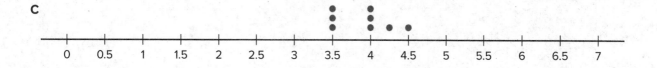

1. One of these data sets has a mean of 5.57 minutes and another has a mean of 3.91 minutes.

   a. Which dot plot shows each of these data sets?

   b. Calculate the mean for the data set on the other dot plot.

2. One of these data sets has a mean absolute deviation of 0.30 and another has a MAD of 0.44.

   a. Which dot plot shows each of these data sets?

   b. Calculate the MAD for the other data set.

3. Do you think the three groups are very different or not? Be prepared to explain your reasoning.

4. A fourth album has a mean length of 8 minutes with a mean absolute deviation of 1.2. Is this data set very different from each of the others?

NAME _____ DATE _____ PERIOD _____

Summary
Comparing Groups

Comparing two individuals is fairly straightforward. The question "Which dog is taller?" can be answered by measuring the heights of two dogs and comparing them directly.

Comparing two groups can be more challenging. What does it mean for the basketball team to generally be taller than the soccer team?

To compare two groups, we use the distribution of values for the two groups. Most importantly, a measure of center (usually **mean** or **median**) and its associated measure of variability (usually **mean absolute deviation** or *interquartile range*) can help determine the differences between groups.

For example, if the average height of pugs in a dog show is 11 inches, and the average height of the beagles in the dog show is 15 inches, it seems that the beagles are generally taller.
On the other hand, if the MAD is 3 inches, it would not be unreasonable to find a beagle that is 11 inches tall or a pug that is 14 inches tall. Therefore the heights of the two dog breeds may not be very different from one another.

Glossary

mean
mean absolute deviation (MAD)
median

Practice
Comparing Groups

1. Compare the weights of the backpacks for the students in these three classes.

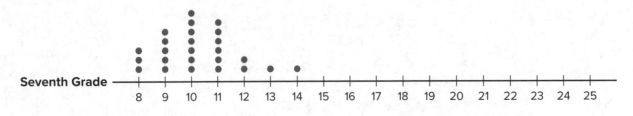

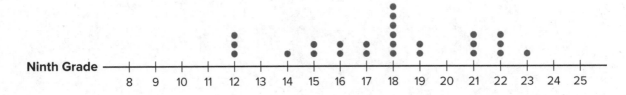

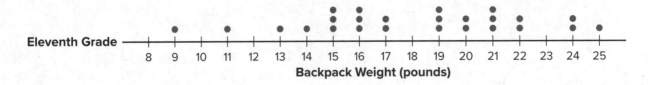

Backpack Weight (pounds)

NAME _____ DATE _____ PERIOD _____

2. A bookstore has marked down the price for all the books in a certain series by 15%. (Lesson 4-11)

   a. How much is the discount on a book that normally costs $18.00?

   b. After the discount, how much would the book cost?

3. Match each expression in the first list with an equivalent expression from the second list. (Lesson 6-22)

   Expressions   **Equivalent Expressions**

   a. $6(x + 2y) - 2(y - 2x)$ i. $10(x - y)$

   b. $2.5(2x + 4y) - 5(4y - x)$ ii. $10(x + y)$

   c. $4(5x - 3y) - 10x + 6y$ iii. $10x + 6y$

   d. $5.5(x + y) - 2(x + y) + 6.5(x + y)$ iv. $10x - 6y$

   e. $7.9(5x + 3y) - 4.2(5x + 3y) - 1.7(5x + 3y)$

4. Angles C and D are complementary. The ratio of the measure of Angle C to the measure of Angle D is $2 : 3$. Find the measure of each angle. Explain or show your reasoning. **(Lesson 7-2)**

Lesson 8-12

Larger Populations

NAME _____ DATE _____ PERIOD _____

Learning Goal Let's compare larger groups.

Warm Up
12.1 First Name versus Last Name

Consider the question: In general, do the students at this school have more letters in their first name or last name? How many more letters?

1. What are some ways you might get some data to answer the question?

2. The other day, we compared the heights of people on different teams and the lengths of songs on different albums. What makes this question about first and last names harder to answer than those questions?

Activity
12.2 John Jacobjingleheimerschmidt

Continue to consider the question from the warm-up: In general, do the students at this school have more letters in their first name or last name? How many more letters?

1. How many letters are in your first name? In your last name?

2. Do the number of letters in your own first and last names give you enough information to make conclusions about students' names in your entire school? Explain your reasoning.

3. Your teacher will provide you with data from the class. Record the mean number of letters as well as the mean absolute deviation for each data set.

   a. The first names of the students in your class.

   b. The last names of the students in your class.

4. Which mean is larger? By how much? What does this difference tell you about the situation?

5. Do the mean numbers of letters in the first and last names for everyone in your class give you enough information to make conclusions about students' names in your entire school? Explain your reasoning.

 Activity
12.3 Siblings and Pets

Consider the question: Do people who are the only child *have more pets*?

1. Earlier, we used information about the people in your class to answer a question about the entire school. Would surveying only the people in your class give you enough information to answer this new question? Explain your reasoning.

2. If you had to have an answer to this question by the end of class today, how would you gather data to answer the question?

3. If you could come back tomorrow with your answer to this question, how would you gather data to answer the question?

4. If someone else in the class came back tomorrow with an answer that was different than yours, what would that mean? How would you determine which answer was better?

NAME _____ DATE _____ PERIOD _____

Activity

12.4 Sampling the Population

For each question, identify the **population** and a possible **sample**.

1. What is the mean number of pages for novels that were on the best seller list in the 1990s?

2. What fraction of new cars sold between August 2010 and October 2016 were built in the United States?

3. What is the median income for teachers in North America?

4. What is the average lifespan of Tasmanian devils?

Are you ready for more?

Political parties often use samples to poll people about important issues. One common method is to call people and ask their opinions. In most places, though, they are not allowed to call cell phones. Explain how this restriction might lead to inaccurate samples of the population.

Summary
Larger Populations

A **population** is a set of people or things that we want to study. Here are some examples of populations:

- All people in the world
- All seventh graders at a school
- All apples grown in the U.S.

A **sample** is a subset of a population. Here are some examples of samples from the listed populations:

- The leaders of each country
- The seventh graders who are in band
- The apples in the school cafeteria

When we want to know more about a population but it is not feasible to collect data from everyone in the population, we often collect data from a sample.

In the lessons that follow, we will learn more about how to pick a sample that can help answer questions about the entire population.

Glossary

population
sample

NAME _____ DATE _____ PERIOD _____

Practice
Larger Populations

1. Suppose you are interested in learning about how much time seventh grade students at your school spend outdoors on a typical school day.

   Select **all** the samples that are a part of the population you are interested in.

   (A.) The 20 students in a seventh grade math class.

   (B.) The first 20 students to arrive at school on a particular day.

   (C.) The seventh grade students participating in a science fair put on by the four middle schools in a school district.

   (D.) The 10 seventh graders on the school soccer team.

   (E.) The students on the school debate team.

2. For each sample given, list two possible populations they could belong to.

   a. Sample: The prices for apples at two stores near your house.

   b. Sample: The days of the week the students in your math class ordered food during the past week.

   c. Sample: The daily high temperatures for the capital cities of all 50 U.S. states over the past year.

3. If 6 coins are flipped, find the probability that there is at least 1 heads.
   (Lesson 8-9)

4. A school's art club holds a bake sale on Fridays to raise money for art
   supplies. Here are the number of cookies they sold each week in the fall
   and in the spring. (Lesson 8-11)

   Fall

20	26	25	24	29	20	19	19
24	24						

   Spring

19	27	29	21	25	22	26	21
25	25						

   a. Find the mean number of cookies sold in the fall and in the spring.

   b. The MAD for the fall data is 2.8 cookies. The MAD for the spring data
      is 2.6 cookies. Express the difference in means as a multiple of the
      larger MAD.

   c. Based on this data, do you think that sales were generally higher in the
      spring than in the fall?

5. A school is selling candles for a fundraiser. They keep 40% of the total
   sales as their commission, and they pay the rest to the candle company.
   How much money must the school pay to the candle company? (Lesson 4-11)

Price of Candle	Number of Candles Sold
small candle: $11	68
medium candle: $18	45
large candle: $25	21

Lesson 8-13

What Makes a Good Sample?

NAME _____ DATE _____ PERIOD _____

Learning Goal Let's see what makes a good sample.

Warm Up
13.1 Number Talk: Division by Powers of 10

Find the value of each quotient mentally.

1. $34,000 \div 10$ 2. $340 \div 100$ 3. $34 \div 10$ 4. $3.4 \div 100$

Activity
13.2 Selling Paintings

Your teacher will assign you to work with either means or medians.

1. A young artist has sold 10 paintings. Calculate the measure of center you were assigned for each of these samples:

   a. The first two paintings she sold were for $50 and $350.

   b. At a gallery show, she sold three paintings for $250, $400, and $1,200.

   c. Her oil paintings have sold for $410, $400, and $375.

2. Here are the selling prices for all 10 of her paintings.

   $50 $200 $250 $275 $280 $350 $375 $400

   $410 $1,200

   Calculate the measure of center you were assigned for all of the selling prices.

3. Compare your answers with your partner. Were the measures of center for any of the samples close to the same measure of center for the population?

Activity

13.3 Sampling the Fish Market

The price per pound of catfish at a fish market was recorded for 100 weeks.

1. Here are dot plots showing the population and three different samples from that population. What do you notice? What do you wonder?

Population

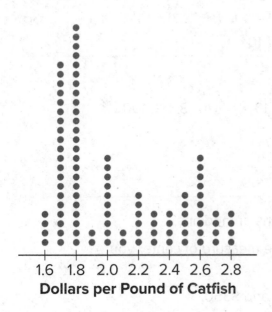

Sample 1

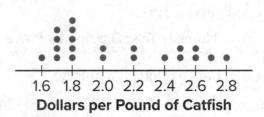

Sample 2

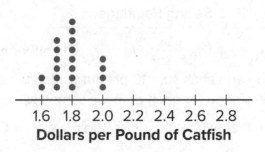

Sample 3

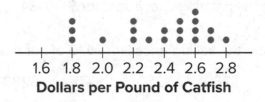

NAME _____ DATE _____ PERIOD _____

2. If the goal is to have the sample represent the population, which of the samples would work best? Which wouldn't work so well? Explain your reasoning.

Are you ready for more?

When doing a statistical study, it is important to keep the goal of the study in mind. Representative samples give us the best information about the distribution of the population as a whole, but sometimes a representative sample won't work for the goal of a study!

For example, suppose you want to study how discrimination affects people in your town. Surveying a representative sample of people in your town would give information about how the population generally feels, but might miss some smaller groups. Describe a way you might choose a sample of people to address this question.

An online shopping company tracks how many items they sell in different categories during each month for a year. Three different auditors each take samples from that data. Use the samples to draw dot plots of what the population data might look like for the furniture and electronics categories.

Auditor 1's Sample

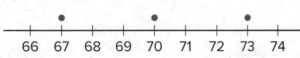

Monthly Sales of Furniture Online (in hundreds)

Auditor 2's Sample

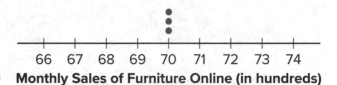

Monthly Sales of Furniture Online (in hundreds)

Auditor 3's Sample

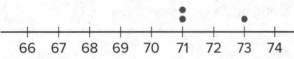

Monthly Sales of Furniture Online (in hundreds)

Population

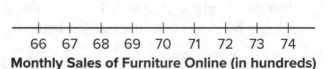

Monthly Sales of Furniture Online (in hundreds)

Auditor 1's Sample

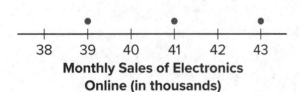

Monthly Sales of Electronics Online (in thousands)

Auditor 2's Sample

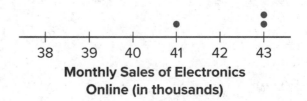

Monthly Sales of Electronics Online (in thousands)

Auditor 3's Sample

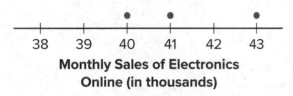

Monthly Sales of Electronics Online (in thousands)

Population

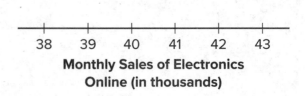

Monthly Sales of Electronics Online (in thousands)

NAME _____ DATE _____ PERIOD _____

Summary
What Makes a Good Sample?

A sample that is **representative** of a population has a distribution that closely resembles the distribution of the population in shape, center, and spread.

For example, consider the distribution of plant heights, in cm, for a population of plants shown in this dot plot. The mean for this population is 4.9 cm, and the MAD is 2.6 cm.

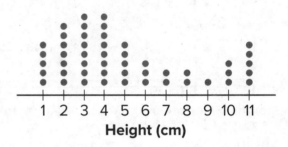

A representative sample of this population should have a larger peak on the left and a smaller one on the right, like this one. The mean for this sample is 4.9 cm, and the MAD is 2.3 cm.

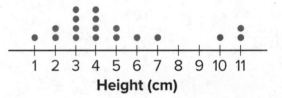

Here is the distribution for another sample from the same population. This sample has a mean of 5.7 cm and a MAD of 1.5 cm. These are both very different from the population, and the distribution has a very different shape, so it is not a representative sample.

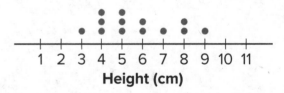

> **Glossary**
>
> **representative**

Practice

What Makes a Good Sample?

1. Suppose 45% of all the students at Andre's school brought in a can of food to contribute to a canned food drive. Andre picks a representative sample of 25 students from the school and determines the sample's percentage.

   He expects the percentage for this sample will be 45%. Do you agree? Explain your reasoning.

2. This is a dot plot of the scores on a video game for a population of 50 teenagers.

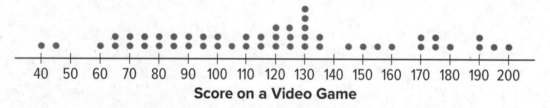

Score on a Video Game

   The three dot plots together are the scores of teenagers in three samples from this population. Which of the three samples is most representative of the population? Explain how you know.

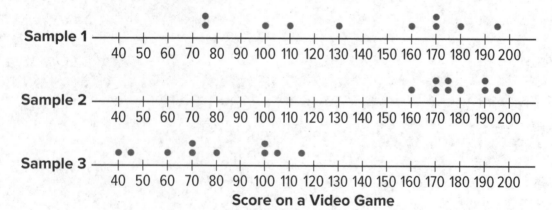

Score on a Video Game

NAME _____ DATE _____ PERIOD _____

3. This is a dot plot of the number of text messages sent one day for a sample of the students at a local high school. The sample consisted of 30 students and was selected to be representative of the population.

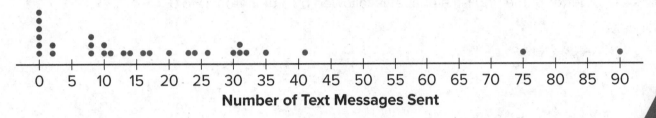

Number of Text Messages Sent

a. What do the six values of 0 in the dot plot represent?

b. Since this sample is representative of the population, describe what you think a dot plot for the entire population might look like.

4. A doctor suspects you might have a certain strain of flu and wants to test your blood for the presence of markers for this strain of virus. Why would it be good for the doctor to take a sample of your blood rather than use the population? **(Lesson 8-12)**

5. How many different outcomes are in each sample space? Explain your reasoning. (Lesson 8-8)

(You do not need to write out the actual options, just provide the number and your reasoning.)

a. A letter of the English alphabet is followed by a digit from 0 to 9.

b. A baseball team's cap is selected from 3 different colors, 2 different clasps, and 4 different locations for the team logo. A decision is made to include or not to include reflective piping.

c. A locker combination like 7-23-11 uses three numbers, each from 1 to 40. Numbers can be used more than once, like 7-23-7.

Lesson 8-14

Sampling in a Fair Way

NAME _____ DATE _____ PERIOD _____

Learning Goal Let's explore ways to get representative samples.

Warm Up
14.1 Ages of Moviegoers

A survey was taken at a movie theater to estimate the average age of moviegoers.

Here is a dot plot showing the ages of the first 20 people surveyed.

1. What questions do you have about the data from survey?

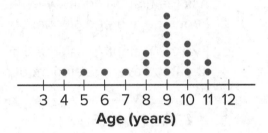

2. What assumptions would you make based on these results?

Activity
14.2 Comparing Methods for Selecting Samples

Take turns with your partner reading each option aloud. For each situation, discuss:

- Would the different methods for selecting a sample lead to different conclusions about the population?

- What are the benefits of each method?

- What might each method overlook?

- Which of the methods listed would be the most likely to produce samples that are representative of the population being studied?

- Can you think of a better way to select a sample for this situation?

1. Lin is running in an election to be president of the seventh grade. She wants to predict her chances of winning. She has the following ideas for surveying a sample of the students who will be voting:

   a. Ask everyone on her basketball team who they are voting for.

   b. Ask every third girl waiting in the lunch line who they are voting for.

   c. Ask the first 15 students to arrive at school one morning who they are voting for.

2. A nutritionist wants to collect data on how much caffeine the average American drinks per day. She has the following ideas for how she could obtain a sample:

   a. Ask the first 20 adults who arrive at a grocery store after 10:00 a.m. about the average amount of caffeine they consume each day.

   b. Every 30 minutes, ask the first adult who comes into a coffee shop about the average amount of caffeine they consume each day.

 Activity

14.3 That's the First Straw

Your teacher will have some students draw straws from a bag.

1. As each straw is taken out and measured, record its length (in inches) in the table.

	Straw 1	Straw 2	Straw 3	Straw 4	Straw 5
Sample 1					
Sample 2					

2. Estimate the mean length of all the straws in the bag based on:

   a. the mean of the first sample.

   b. the mean of the second sample.

NAME _____ DATE _____ PERIOD _____

3. Were your two estimates the same? Did the mean length of all the straws in the bag change in between selecting the two samples? Explain your reasoning.

4. The actual mean length of all of the straws in the bag is about 2.37 inches. How do your estimates compare to this mean length?

5. If you repeated the same process again but you selected a larger sample (such as 10 or 20 straws, instead of just 5), would your estimate be more accurate? Explain your reasoning.

Activity
14.4 That's the Last Straw

There were a total of 35 straws in the bag. Suppose we put the straws in order from shortest to longest and then assigned each straw a number from 1 to 35. For each of these methods, decide whether it would be fair way to select a sample of 5 straws. Explain your reasoning.

1. Select the straws numbered 1 through 5.

2. Write the numbers 1 through 35 on pieces of paper that are all the same size. Put the papers into a bag. Without looking, select five papers from the bag. Use the straws with those numbers for your sample.

3. Using the same bag as the previous question, select one paper from the bag. Use the number on that paper to select the first straw for your sample. Then use the next 4 numbers in order to complete your sample. (For example, if you select number 17, then you also use straws 18, 19, 20, and 21 for your sample.)

4. Create a spinner with 35 sections that are all the same size, and number them 1 through 35. Spin the spinner 5 times and use the straws with those numbers for your sample.

Are you ready for more?

Computers accept inputs, follow instructions, and produce outputs, so they cannot produce truly random numbers. If you knew the input, you could predict the output by following the same instructions the computer is following. When truly random numbers are needed, scientists measure natural phenomena such as radioactive decay or temperature variations. Before such measurements were possible, statisticians used random number tables, like this:

85 67 95 02 42 61 21 35 15 34 41
85 94 61 72 53 24 15 67 85 94 12
67 88 15 32 42 65 75 98 46 25 13
07 53 60 75 82 34 67 44 20 42 33
99 37 40 33 40 88 90 50 75 22 90
00 03 84 57 91 15 70 08 90 03 02
78 07 16 51 13 89 67 64 54 05 26
62 06 61 43 02 60 73 58 38 53 88
02 50 88 44 37 05 13 54 78 97 30

Use this table to select a sample of 5 straws. Pick a starting point at random in the table. If the number is between 01 and 35, include that number straw in your sample. If the number has already been selected, or is not between 01 and 35, ignore it, and move on to the next number.

NAME _____ DATE _____ PERIOD _____

Summary
Sampling in a Fair Way

A sample is *selected at random* from a population if it has an equal chance of being selected as every other sample of the same size.

For example, if there are 25 students in a class, then we can write each of the students' names on a slip of paper and select 5 papers from a bag to get a sample of 5 students selected at random from the class.

Other methods of selecting a sample from a population are likely to be *biased*. This means that it is less likely that the sample will be representative of the population as a whole.

For example, if we select the first 5 students who walk in the door, that will not give us a random sample because students who typically come late are not likely to be selected.

- A sample that is selected at random may not always be a representative sample, but it is more likely to be representative than using other methods.

- It is not always possible to select a sample at random. For example, if we want to know the average length of wild salmon, it is not possible to identify each one individually, select a few at random from the list, and then capture and measure those exact fish.

- When a sample cannot be selected at random, it is important to try to reduce bias as much as possible when selecting the sample.

Practice
Sampling in a Fair Way

1. The meat department manager at a grocery store is worried some of the packages of ground beef labeled as having one pound of meat may be under-filled. He decides to take a sample of 5 packages from a shipment containing 100 packages of ground beef. The packages were numbered as they were put in the box, so each one has a different number between 1 and 100.

   Describe how the manager can select a fair sample of 5 packages.

2. Select **all** the reasons why random samples are preferred over other methods of getting a sample.

   (A.) If you select a random sample, you can determine how many people you want in the sample.

   (B.) A random sample is always the easiest way to select a sample from a population.

   (C.) A random sample is likely to give you a sample that is representative of the population.

   (D.) A random sample is a fair way to select a sample, because each person in the population has an equal chance of being selected.

   (E.) If you use a random sample, the sample mean will always be the same as the population mean.

NAME _____ DATE _____ PERIOD _____

3. Jada is using a computer's random number generator to produce 6 random whole numbers between 1 and 100 so she can use a random sample. The computer produces the numbers: 1, 2, 3, 4, 5, and 6. Should she use these numbers or have the computer generate a new set of random numbers? Explain your reasoning.

4. A group of 100 people is divided into 5 groups with 20 people in each. One person's name is chosen, and everyone in their group wins a prize. Noah simulates this situation by writing 100 different names on papers and putting them in a bag, then drawing one out. Kiran suggests there is a way to do it with fewer paper slips. Explain a method that would simulate this situation with fewer than 100 slips of paper. **(Lesson 8-6)**

5. Data collected from a survey of American teenagers aged 13 to 17 was used to estimate that 29% of teens believe in ghosts. This estimate was based on data from 510 American teenagers. What is the population that people carrying out the survey were interested in? **(Lesson 8-12)**

A. All people in the United States.

B. The 510 teens that were surveyed.

C. All American teens who are between the ages of 13 and 17.

D. The 29% of the teens surveyed who said they believe in ghosts.

6. A computer simulates flipping a coin 100 times, then counts the longest string of heads in a row.

   Based on these results, estimate the probability that there will be at least 15 heads in a row. (Lesson 8-7)

Trial	Most Heads in a Row
1	8
2	6
3	5
4	11
5	13

Lesson 8-15

Estimating Population Measures of Center

NAME _____ DATE _____ PERIOD _____

Learning Goal Let's use samples to estimate measures of center for the population.

 ## Warm Up
15.1 Describing the Center

Would you use the median or mean to describe the center of each data set? Explain your reasoning.

Heights of 50 Basketball Players

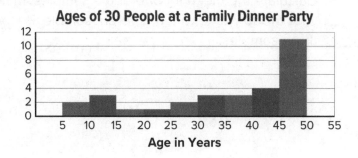

Ages of 30 People at a Family Dinner Party

Backpack Weights of 6th-Grade Students

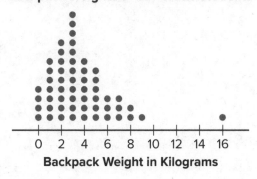

How Many Books Students Read Over Summer Break

Activity

15.2 Three Different TV Shows

Here are the ages (in years) of a random sample of 10 viewers for 3 different television shows. The shows are titled, "Science Experiments YOU Can Do," "Learning to Read," and "Trivia the Game Show."

Sample 1 6 6 5 4 8 5 7 8 6 6

Sample 2 15 14 12 13 12 10 12 11 10 8

Sample 3 43 60 50 36 58 50 73 59 69 51

1. Calculate the mean for *one* of the samples. Make sure each person in your group works with a different sample. Record the answers for all three samples.

2. Which show do you think each sample represents? Explain your reasoning.

NAME _____ DATE _____ PERIOD _____

Activity

15.3 Who's Watching What?

Here are three more samples of viewer ages collected for these same 3 television shows.

Sample 4   57 71 5   54 52 13 59 65 10 71

Sample 5   15 5 4 5 4 3 25 2 8 3

Sample 6   6 11 9 56 1 3 11 10 11 2

1. Calculate the mean for *one* of these samples. Record all three answers.

2. Which show do you think each of these samples represents? Explain your reasoning.

3. For each show, estimate the mean age for all the show's viewers.

4. Calculate the mean absolute deviation for *one* of the shows' samples. Make sure each person in your group works with a different sample. Record all three answers.

	Learning to Read	Science Experiments YOU Can Do	Trivia the Game Show
Which sample?			
MAD			

5. What do the different values for the MAD tell you about each group?

6. An advertiser has a commercial that appeals to 15- to 16-year-olds. Based on these samples, are any of these shows a good fit for this commercial? Explain or show your reasoning.

Activity

15.4 Movie Reviews

A movie rating website has many people rate a new movie on a scale of 0 to 100. Here is a dot plot showing a random sample of 20 of these reviews.

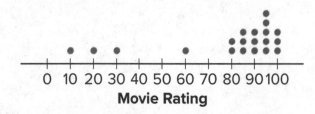

Movie Rating

NAME _____ DATE _____ PERIOD _____

1. Would the mean or median be a better measure for the center of this data? Explain your reasoning.

2. Use the sample to estimate the measure of center that you chose for *all* the reviews.

3. For this sample, the mean absolute deviation is 19.6, and the interquartile range is 15. Which of these values is associated with the measure of center that you chose?

4. Movies must have an average rating of 75 or more from all the reviews on the website to be considered for an award. Do you think this movie will be considered for the award? Use the measure of center and measure of variability that you chose to justify your answer.

Are you ready for more?

Estimate typical temperatures in the United States today by looking up current temperatures in several places across the country. Use the data you collect to decide on the appropriate measure of center for the country, and calculate the related measure of variation for your sample.

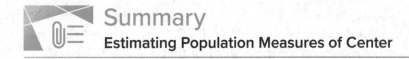

Summary
Estimating Population Measures of Center

Some populations have greater variability than others. For example, we would expect greater variability in the weights of dogs at a dog park than at a beagle meetup.

Dog Park
Mean weight: 12.8 kg
MAD: 2.3 kg

Beagle Meetup
Mean weight: 10.1 kg
MAD: 0.8 kg

The lower MAD indicates there is less variability in the weights of the beagles.

We would expect that the mean weight from a sample that is randomly selected from a group of beagles will provide a more accurate estimate of the mean weight of all the beagles than a sample of the same size from the dogs at the dog park.

In general, a sample of a similar size from a population with *less* variability is *more likely* to have a mean that is close to the population mean.

Glossary

interquartile range

NAME _____ DATE _____ PERIOD _____

 Practice
Estimating Population Measures of Center

1. A random sample of 15 items were selected.

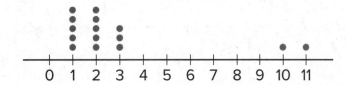

For this data set, is the mean or median a better measure of center? Explain your reasoning.

2. A video game developer wants to know how long it takes people to finish playing their new game. They surveyed a random sample of 13 players and asked how long it took them (in minutes).

| 1,235 | 952 | 457 | 1,486 | 1,759 | 1,148 | 548 | 1,037 |

| 1,864 | 1,245 | 976 | 866 | 1,431 |

a. Estimate the median time it will take *all* players to finish this game.

b. Find the interquartile range for this sample.

3. Han and Priya want to know the mean height of the 30 students in their dance class. They each select a random sample of 5 students.

- The mean height for Han's sample is 59 inches.

- The mean height for Priya's sample is 61 inches.

Does it surprise you that the two sample means are different? Are the population means different? Explain your reasoning.

4. Clare and Priya each took a random sample of 25 students at their school.

- Clare asked each student in her sample how much time they spend doing homework each night. The sample mean was 1.2 hours and the MAD was 0.6 hours.

- Priya asked each student in her sample how much time they spend watching TV each night. The sample mean was 2 hours and the MAD was 1.3 hours.

a. At their school, do you think there is more variability in how much time students spend doing homework or watching TV? Explain your reasoning.

b. Clare estimates the students at her school spend an average of 1.2 hours each night doing homework. Priya estimates the students at her school spend an average of 2 hours each night watching TV. Which of these two estimates is likely to be closer to the actual mean value for all the students at their school? Explain your reasoning.

Lesson 8-16

Estimating Population Proportions

NAME _____ DATE _____ PERIOD _____

Learning Goal Let's estimate population proportions using samples.

 ## Warm Up
16.1 Getting to School

A teacher asked all the students in one class how many minutes it takes them to get to school. Here is a list of their responses:

20	10	15	8	5	15	10	5
20	5	15	10	3	10	18	5
25	5	5	12	10	30	5	10

1. What fraction of the students in this class say:

   a. it takes them 5 minutes to get to school?

   b. it takes them more than 10 minutes to get to school?

2. If the whole school has 720 students, can you use this data to estimate how many of them would say that it takes them more than 10 minutes to get to school?

   Be prepared to explain your reasoning.

Activity

16.2 Reaction Times

The track coach at a high school needs a student whose reaction time is less than 0.4 seconds to help out at track meets. All the twelfth graders in the school measured their reaction times. Your teacher will give you a bag of papers that list their results.

1. Work with your partner to select a random sample of 20 reaction times, and record them in the table.

2. What **proportion** of your sample is less than 0.4 seconds?

3. Estimate the proportion of all twelfth graders at this school who have a reaction time of less than 0.4 seconds. Explain your reasoning.

4. There are 120 twelfth graders at this school. Estimate how many of them have a reaction time of less than 0.4 seconds.

5. Suppose another group in your class comes up with a different estimate than yours for the previous question.

    a. What is another estimate that would be *reasonable*?

    b. What is an estimate you would consider *unreasonable*?

NAME _____ DATE _____ PERIOD _____

Activity

16.3 A New Comic Book Hero

Here are the results of a survey of 20 people who read *The Adventures of Super Sam* regarding what special ability they think the new hero should have.

Response	What New Ability?
1	fly
2	freeze
3	freeze
4	fly
5	fly
6	freeze
7	fly
8	super strength
9	freeze
10	fly

Response	What New Ability?
11	freeze
12	freeze
13	fly
14	invisibility
15	freeze
16	fly
17	freeze
18	fly
19	super strength
20	freeze

1. What proportion of this sample want the new hero to have the ability to fly?

2. If there are 2,024 dedicated readers of *The Adventures of Super Sam*, estimate the number of readers who want the new hero to fly.

Two other comic books did a similar survey of their readers.

- In a survey of people who read *Beyond Human*, 42 out of 60 people want a new hero to be able to fly.

- In a survey of people who read *Mysterious Planets*, 14 out of 40 people want a new hero to be able to fly.

3. Do you think the proportion of all readers who want a new hero that can fly are nearly the same for the three different comic books? Explain your reasoning.

4. If you were in charge of these three comics, would you give the ability to fly to any of the new heroes? Explain your reasoning using the proportions you calculated.

Activity

16.4 Flying to the Shelves

The authors of *The Adventures of Super Sam* chose 50 different random samples of readers. Each sample was of size 20. They looked at the sample proportions who prefer the new hero to fly.

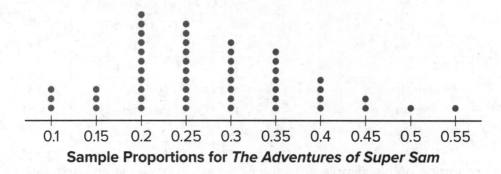

Sample Proportions for *The Adventures of Super Sam*

1. What is a good estimate of the proportion of *all* readers who want the new hero to be able to fly?

2. Are most of the sample proportions within 0.1 of your estimate for the population proportion?

3. If the authors of *The Adventures of Super Sam* give the new hero the ability to fly, will that please most of the readers? Explain your reasoning.

NAME _____ DATE _____ PERIOD _____

The authors of the other comic book series created similar dot plots.

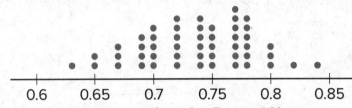

Sample Proportions for *Beyond Human*

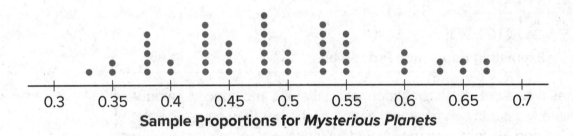

Sample Proportions for *Mysterious Planets*

4. For each of these series, estimate the proportion of all readers who want the new hero to fly.

   a. *Beyond Human*:

   b. *Mysterious Planets*:

5. Should the authors of either of these series give their new hero the ability to fly?

6. Why might it be more difficult for the authors of *Mysterious Planets* to make the decision than the authors of the other series?

Draw an example of a dot plot with at least 20 dots that represent the sample proportions for different random samples that would indicate that the population proportion is above 0.6, but there is a lot of uncertainty about that estimate.

Summary
Estimating Population Proportions

Sometimes a data set consists of information that fits into specific categories. For example, we could survey students about whether they have a pet cat or dog. The categories for these data would be {neither, dog only, cat only, both}. Suppose we surveyed 10 students. Here is a table showing possible results:

Option	Number of Responses
Neither dog nor cat	2
Dog only	4
Cat only	1
Both dog and cat	3

In this sample, 3 of the students said they have both a dog and a cat. We can say that the **proportion** of these students who have a both a dog and a cat is $\frac{3}{10}$ or 0.3.

If this sample is representative of all 720 students at the school, we can predict that about $\frac{3}{10}$ of 720, or about 216 students at the school have both a dog and a cat.

In general, a proportion is a number from 0 to 1 that represents the fraction of the data that belongs to a given category.

Glossary

proportion

NAME _____ DATE _____ PERIOD _____

Practice
Estimating Population Proportions

1. Tyler wonders what proportion of students at his school would dye their hair blue, if they were allowed to. He surveyed a random sample of 10 students at his school, and 2 of them said they would. Kiran didn't think Tyler's estimate was very accurate, so he surveyed a random sample of 100 students, and 17 of them said they would.

   a. Based on Tyler's sample, estimate what proportion of the students would dye their hair blue.

   b. Based on Kiran's sample, estimate what proportion of the students would dye their hair blue.

   c. Whose estimate is more accurate? Explain how you know.

2. Han surveys a random sample of students about their favorite pasta dish served by the cafeteria and makes a bar graph of the results.

   Estimate the proportion of the students who like lasagna as their favorite pasta dish.

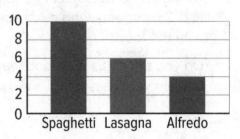

3. Elena wants to know what proportion of people have cats as pets. Describe a process she could use to estimate an answer to her question.

4. The science teacher gives daily homework. For a random sample of days throughout the year, the median number of problems is 5 and the IQR is 2. The Spanish teacher also gives daily homework. For a random sample of days throughout the year, the median number of problems is 10 and the IQR is 1. If you estimate the median number of science homework problems to be 5 and the median number of Spanish problems to be 10, which is more likely to be accurate? Explain your reasoning. (Lesson 8-15)

5. Diego wants to survey a sample of students at his school to learn about the percentage of students who are satisfied with the food in the cafeteria. He decides to go to the cafeteria on a Monday and ask the first 25 students who purchase a lunch at the cafeteria if they are satisfied with the food.

   Do you think this is a good way for Diego to select his sample? Explain your reasoning. (Lesson 8-14)

Lesson 8-17

More about Sampling Variability

NAME _____ DATE _____ PERIOD _____

Learning Goal Let's compare samples from the same population.

Warm Up
17.1 Average Reactions

The other day, you worked with the reaction times of twelfth graders to see if they were fast enough to help out at the track meet. Look back at the sample you collected.

1. Calculate the mean reaction time for your sample.

2. Did you and your partner get the same sample mean? Explain why or why not.

Activity
17.2 Reaction Population

Your teacher will display a blank dot plot.

1. Plot your sample mean from the previous activity on your teacher's dot plot.

2. What do you notice about the distribution of the sample means from the class?

   a. Where is the center?

   b. Is there a lot of variability?

   c. Is it approximately symmetric?

3. The population mean is 0.442 seconds. How does this value compare to the sample means from the class?

Pause here so your teacher can display a dot plot of the population of reaction times.

4. What do you notice about the distribution of the population?

   a. Where is the center?

   b. Is there a lot of variability?

   c. Is it approximately symmetric?

5. Compare the two displayed dot plots.

6. Based on the distribution of sample means from the class, do you think the mean of a random sample of 20 items is likely to be:

   a. within 0.01 seconds of the actual population mean?

   b. within 0.1 seconds of the actual population mean?

   Explain or show your reasoning.

 Activity

17.3 How Much Do You Trust the Answer?

The other day you worked with 2 different samples of viewers from each of 3 different television shows. Each sample included 10 viewers. Here are the mean ages for 100 different samples of viewers from each show.

NAME _____ DATE _____ PERIOD _____

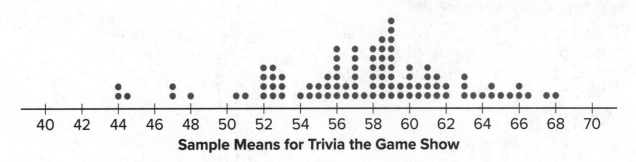

Sample Means for Trivia the Game Show

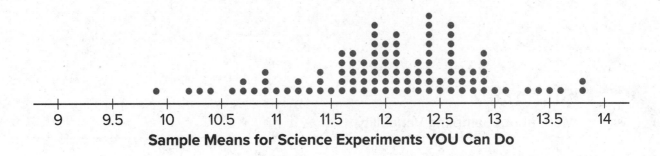

Sample Means for Science Experiments YOU Can Do

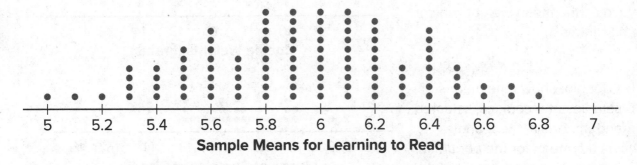

Sample Means for Learning to Read

1. For each show, use the dot plot to estimate the *population* mean.

   a. Trivia the Game Show

   b. Science Experiments YOU Can Do

   c. Learning to Read

2. For each show, are most of the sample means within 1 year of your estimated population mean?

3. Suppose you take a new random sample of 10 viewers for each of the 3 shows. Which show do you expect to have the new sample mean closest to the population mean? Explain or show your reasoning.

Market research shows that advertisements for retirement plans appeal to people between the ages of 40 and 55. Younger people are usually not interested and older people often already have a plan. Is it a good idea to advertise retirement plans during any of these three shows? Explain your reasoning.

Summary

More about Sampling Variability

This dot plot shows the weights, in grams, of 18 cookies. The triangle indicates the mean weight, which is 11.6 grams.

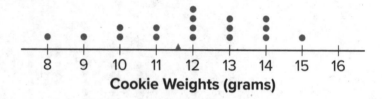

This dot plot shows the *means* of 20 samples of 5 cookies, selected at random. Again, the triangle shows the mean for the *population* of cookies. Notice that most of the sample means are fairly close to the mean of the entire population.

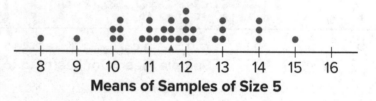

This dot plot shows the means of 20 samples of 10 cookies, selected at random. Notice that the means for these samples are even closer to the mean for the entire population.

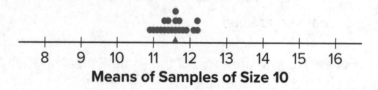

In general, as the sample size gets bigger, the mean of a sample is more likely to be closer to the mean of the population.

NAME _____ DATE _____ PERIOD _____

Practice
More about Sampling Variability

1. One thousand baseball fans were asked how far they would be willing to travel to watch a professional baseball game. From this population, 100 different samples of size 40 were selected. Here is a dot plot showing the mean of each sample.

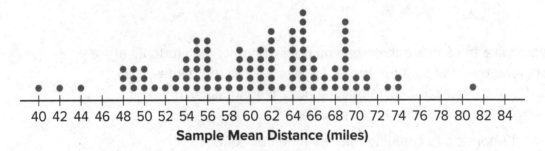

Sample Mean Distance (miles)

Based on the distribution of sample means, what do you think is a reasonable estimate for the mean of the population?

2. Last night, everyone at the school music concert wrote their age on a slip of paper and placed it in a box. Today, each of the students in a math class selected a random sample of size 10 from the box of papers. Here is a dot plot showing their sample means, rounded to the nearest year.

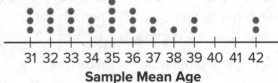

Sample Mean Age

a. Does the number of dots on the dot plot tell you how many people were at the concert or how many students are in the math class?

b. The mean age for the population was 35 years. If Elena picks a new sample of size 10 from this population, should she expect her sample mean to be within 1 year of the population mean? Explain your reasoning.

c. What could Elena do to select a random sample that is more likely to have a sample mean within 1 year of the population mean?

3. A random sample of people were asked which hand they prefer to write with. "L" means they prefer to use their left hand, and "R" means they prefer to use their right hand.

Here are the results.

L R R R R R R

R R L R R R R

Based on this sample, estimate the proportion of the population that prefers to write with their left hand. (Lesson 8-16)

4. Andre would like to estimate the mean number of books the students at his school read over the summer break. He has a list of the names of all the students at the school, but he doesn't have time to ask every student how many books they read. (Lesson 8-15)

What should Andre do to estimate the mean number of books?

5. A hockey team has a 75% chance of winning against the opposing team in each game of a playoff series. To win the series, the team must be the first to win 4 games. (Lesson 8-10)

a. Design a simulation for this event.

b. What counts as a successful outcome in your simulation?

c. Estimate the probability using your simulation.

Lesson 8-18

Comparing Populations using Samples

NAME _____ DATE _____ PERIOD _____

Learning Goal Let's compare different populations using samples.

 Warm Up

18.1 Same Mean? Same MAD?

Without calculating, tell whether each pair of data sets have the same mean and whether they have the same mean absolute deviation.

1. **Set A**   1   3   3   5   6   8   10   14

   Set B   21   23   23   25   26   28   30   34

2. **Set X**   1   2   3   4   5

   Set Y   1   2   3   4   5   6

3. **Set P**   47   53   58   62

   Set Q   37   43   68   72

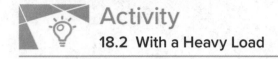

Activity

18.2 With a Heavy Load

Consider the question: Do tenth-grade students' backpacks generally weigh more than seventh-grade students' backpacks?

Here are dot plots showing the weights of backpacks for a random sample of students from these two grades.

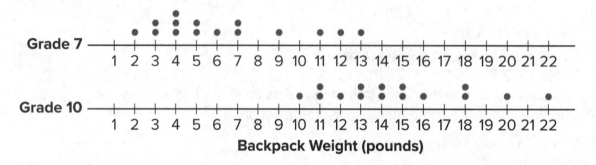

1. Did any seventh-grade backpacks in this sample weigh more than a tenth-grade backpack?

2. The mean weight of this sample of seventh-grade backpacks is 6.3 pounds. Do you think the mean weight of backpacks for *all* seventh-grade students is exactly 6.3 pounds?

3. The mean weight of this sample of tenth-grade backpacks is 14.8 pounds. Do you think there is a meaningful difference between the weight of all seventh-grade and tenth-grade students' backpacks? Explain or show your reasoning.

NAME _____ DATE _____ PERIOD _____

Activity

18.3 Do They Carry More?

Here are 10 more random samples of seventh-grade students' backpack weights.

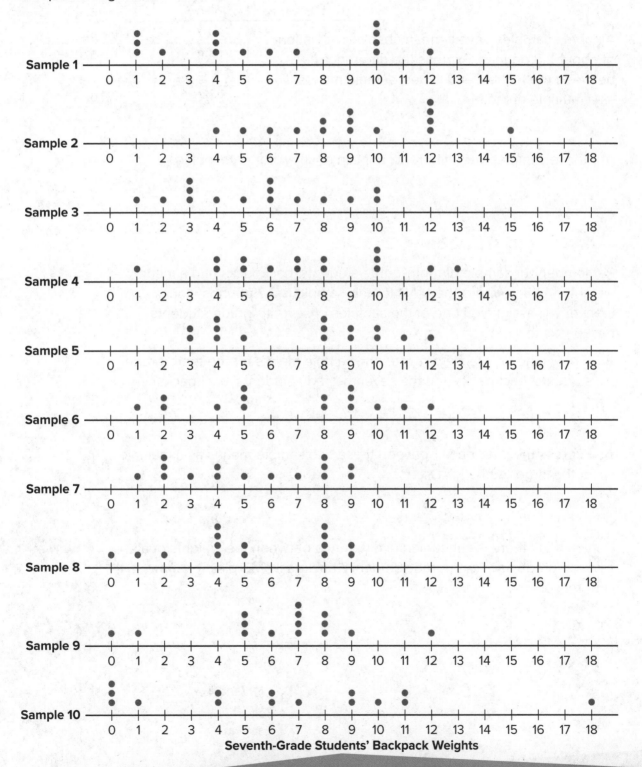

Seventh-Grade Students' Backpack Weights

1. a. Which sample has the highest mean weight?

   b. Which sample has the lowest mean weight?

   c. What is the difference between these two sample means?

Sample Number	Mean Weight (pounds)
1	5.8
2	9.2
3	5.5
4	7.3
5	7.2
6	6.6
7	5.2
8	5.3
9	6.3
10	6.4

2. All of the samples have a mean absolute deviation of about 2.8 pounds. Express the difference between the highest and lowest sample means as a multiple of the MAD.

3. Are these samples very different? Explain or show your reasoning.

4. Remember our sample of tenth-grade students' backpacks had a mean weight of 14.8 pounds. The MAD for this sample is 2.7 pounds. Your teacher will assign you one of the samples of seventh-grade students' backpacks to use.

   a. What is the difference between the sample means for the tenth-grade students' backpacks and the seventh-grade students' backpacks?

   b. Express the difference between these two sample means as a multiple of the larger of the MADs.

5. Do you think there is a meaningful difference between the weights of all seventh-grade and tenth-grade students' backpacks? Explain or show your reasoning.

NAME _____ DATE _____ PERIOD _____

Activity

18.4 Steel from Different Regions

When anthropologists find steel artifacts, they can test the amount of carbon in the steel to learn about the people that made the artifacts. Here are some box plots showing the percentage of carbon in samples of steel that were found in two different regions.

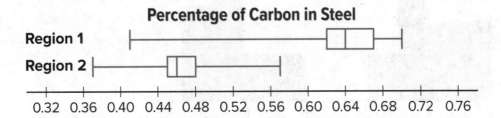

1. Was there any steel found in region 1 that had:

   a. *more* carbon than some of the steel found in region 2?

   b. *less* carbon than some of the steel found in region 2?

2. Do you think there is a meaningful difference between all the steel artifacts found in regions 1 and 2?

3. Which sample has a distribution that is *not* approximately symmetric?

4. What is the difference between the sample medians for these two regions?

	Sample Median (%)	IQR (%)
Region 1	0.64	0.05
Region 2	0.47	0.03

5. Express the difference between these two sample medians as a multiple of the larger interquartile range.

6. The anthropologists who conducted the study concluded that there was a meaningful difference between the steel from these regions. Do you agree? Explain or show your reasoning.

Sometimes we want to compare two different populations. For example, is there a meaningful difference between the weights of pugs and beagles? Here are histograms showing the weights for a sample of dogs from each of these breeds:

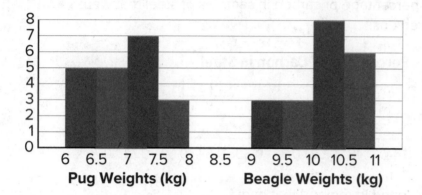

Pug Weights (kg)   **Beagle Weights (kg)**

The red triangles show the mean weight of each sample, 6.9 kg for the pugs and 10.1 kg for the beagles. The red lines show the weights that are within 1 MAD of the mean. We can think of these as "typical" weights for the breed. These typical weights do not overlap. In fact, the distance between the means is 10.1 − 6.9 or 3.2 kg, over 6 times the larger MAD! So we can say there *is* a meaningful difference between the weights of pugs and beagles.

Is there a meaningful difference between the weights of male pugs and female pugs? Here are box plots showing the weights for a sample of male and female pugs:

Male Pug Weights (kg)

Female Pug Weights (kg)

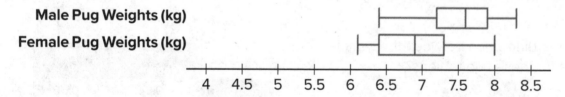

We can see that the medians are different, but the weights between the first and third quartiles overlap. Based on these samples, we would say there is *not* a meaningful difference between the weights of male pugs and female pugs.

In general, if the measures of center for two samples are at least two measures of variability apart, we say the difference in the measures of center is meaningful. Visually, this means the range of typical values does not overlap. If they are closer, then we don't consider the difference to be meaningful.

NAME _____ DATE _____ PERIOD _____

 Practice

Comparing Populations using Samples

1. Lin wants to know if students in elementary school generally spend more time playing outdoors than students in middle school. She selects a random sample of size 20 from each population of students and asks them how many hours they played outdoors last week. Suppose that the MAD for each of her samples is about 3 hours.

   Select **all** pairs of sample means for which Lin could conclude there is a meaningful difference between the two populations.

   (A.) elementary school: 12 hours, middle school: 10 hours

   (B.) elementary school: 14 hours, middle school: 9 hours

   (C.) elementary school: 13 hours, middle school: 6 hours

   (D.) elementary school: 13 hours, middle school: 10 hours

   (E.) elementary school: 7 hours, middle school: 15 hours

2. These two box plots show the distances of a standing jump, in inches, for a random sample of 10-year-olds and a random sample of 15-year-olds.

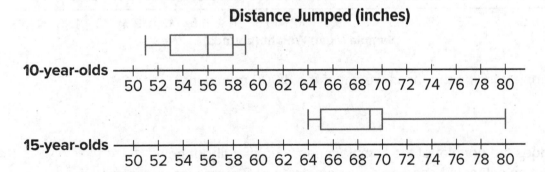

Distance Jumped (inches)

10-year-olds

50 52 54 56 58 60 62 64 66 68 70 72 74 76 78 80

15-year-olds

50 52 54 56 58 60 62 64 66 68 70 72 74 76 78 80

   Is there a meaningful difference in median distance for the two populations? Explain how you know.

3. The median income for a sample of people from Chicago is about $60,000 and the median income for a sample of people from Kansas City is about $46,000, but researchers have determined there is not a meaningful difference in the medians. Explain why the researchers might be correct.

4. A farmer grows 5,000 pumpkins each year. The pumpkins are priced according to their weight, so the farmer would like to estimate the mean weight of the pumpkins he grew this year. He randomly selects 8 pumpkins and weighs them. Here are the weights (in pounds) of these pumpkins:

    2.9   6.8   7.3   7.7   8.9   10.6   12.3   15.3

   a. Estimate the mean weight of the pumpkins the farmer grew. (Lesson 8-17)

   This dot plot shows the mean weight of 100 samples of eight pumpkins, similar to the one above.

   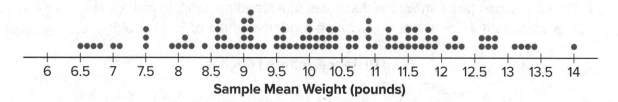

   Sample Mean Weight (pounds)

   b. What appears to be the mean weight of the 5,000 pumpkins?

   c. What does the dot plot of the sample means suggest about how accurate an estimate based on a single sample of 8 pumpkins might be?

   d. What do you think the farmer might do to get a more accurate estimate of the population mean?

Lesson 8-19

Comparing Populations with Friends

NAME _____ DATE _____ PERIOD _____

Learning Goal Let's ask important questions to compare groups.

Warm Up
19.1 Features of Graphic Representations

Dot plots, histograms, and box plots are different ways to represent a data set graphically.

Which of those displays would be the easiest to use to find each feature of the data?

1. the mean

2. the median

3. the mean absolute deviation

4. the interquartile range

5. the symmetry

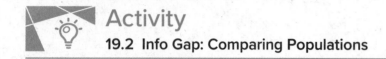

Activity

19.2 Info Gap: Comparing Populations

Your teacher will give you either a *problem card* or a *data card*. Do not show or read your card to your partner.

If your teacher gives you the *problem card*:	If your teacher gives you the *data card*:
1. Silently read your card and think about what information you need to be able to answer the question. 2. Ask your partner for the specific information that you need. 3. Explain how you are using the information to solve the problem. Continue to ask questions until you have enough information to solve the problem. 4. Share the *problem card* and solve the problem independently. 5. Read the *data card* and discuss your reasoning.	1. Silently read your card. 2. Ask your partner *"What specific information do you need?"* and wait for them to *ask* for information. If your partner asks for information that is not on the card, do not do the calculations for them. Tell them you don't have that information. 3. Before sharing the information, ask *"Why do you need that information?"* Listen to your partner's reasoning and ask clarifying questions. 4. Read the *problem card* and solve the problem independently. 5. Share the *data card* and discuss your reasoning.

Pause here so your teacher can review your work. Ask your teacher for a new set of cards and repeat the activity, trading roles with your partner.

Are you ready for more?

Is there a meaningful difference between top sports performance in two different decades? Choose a variable from your favorite sport (for example, home runs in baseball, kills in volleyball, aces in tennis, saves in soccer, etc.) and compare the leaders for each year of two different decades. Is the performance in one decade meaningfully different from the other?

NAME _____ DATE _____ PERIOD _____

Activity
19.3 Comparing to Known Characteristics

1. A college graduate is considering two different companies to apply to for a job. Acme Corp lists this sample of salaries on their website.

   $45,000 $55,000 $140,000 $70,000 $60,000 $50,000

   What typical salary would Summit Systems need to have to be meaningfully different from Acme Corp? Explain your reasoning.

2. A factory manager is wondering whether they should upgrade their equipment. The manager keeps track of how many faulty products are created each day for a week.

   6   7   8   6   7   5   7

   The new equipment guarantees an average of 4 or fewer faulty products per day. Is there a meaningful difference between the new and old equipment? Explain your reasoning.

Summary
Comparing Populations with Friends

When using samples to compare two populations, there are a lot of factors to consider.

- Are the samples representative of their populations? If the sample is biased, then it may not have the same center and variability as the population.

- Which characteristic of the populations makes sense to compare—the mean, the median, or a proportion?

- How variable is the data? If the data is very spread out, it can be more difficult to make conclusions with certainty.

Knowing the correct questions to ask when trying to compare groups is important to correctly interpret the results.

NAME _____ DATE _____ PERIOD _____

 Practice
Comparing Populations with Friends

1. An agent at an advertising agency asks a random sample of people how many episodes of a TV show they watch each day. The results are shown in the dot plot.

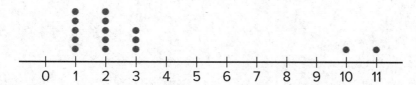

The agency currently advertises on a different show, but wants to change to this one as long as the typical number of episodes is not meaningfully less.

a. What measure of center and measure of variation would the agent need to find for their current show to determine if there is a meaningful difference? Explain your reasoning.

b. What are the values for these same characteristics for the data in the dot plot?

c. What numbers for these characteristics would be meaningfully different if the measure of variability for the current show is similar? Explain your reasoning.

2. Jada wants to know if there is a meaningful difference in the mean number of friends on social media for teens and adults. She looks at the friend count for the 10 most popular of her friends and the friend count for 10 of her parents' friends. She then computes the mean and MAD of each sample and determines there is a meaningful difference. Jada's dad later tells her he thinks she has not come to the right conclusion. Jada checks her calculations and everything is right. Do you agree with her dad? Explain your reasoning.

3. The mean weight for a sample of a certain kind of ring made from platinum is 8.21 grams. The mean weight for a sample of a certain kind of ring made from gold is 8.61 grams. Is there a meaningful difference in the weights of the two types of rings? Explain your reasoning.

4. The lengths in feet of a random sample of 20 male and 20 female humpback whales were measured and used to create the box plot. (Lesson 8-15)

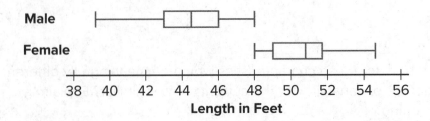

Estimate the median lengths of male and female humpback whales based on these samples.

Lesson 8-20

Memory Test

NAME _____ DATE _____ PERIOD _____

Learning Goal Let's put it all together.

 Activity

20.1 Collecting a Sample

You teacher will give you a paper that lists a data set with 100 numbers in it. Explain whether each method of obtaining a sample of size 20 would produce a random sample.

- Option 1: A spinner has 10 equal sections on it. Spin once to get the row number and again to get the column number for each member of your sample. Repeat this 20 times.

- Option 2: Since the data looks random already, use the first two rows.

- Option 3: Cut up the data and put them into a bag. Shake the bag to mix up the papers, and take out 20 values.

- Option 4: Close your eyes and point to one of the numbers to use as your first value in your sample. Then, keep moving one square from where your finger is to get a path of 20 values for your sample.

 Activity

20.2 Sample Probabilities

Continue working with the data set your teacher gave you in the previous activity. The data marked with a star all came from students at Springfield Middle School.

1. When you select the first value for your random sample, what is the probability that it will be a value that came from a student at Springfield Middle School?

2. What proportion of your entire sample would you expect to be from Springfield Middle School?

3. If you take a random sample of size 10, how many scores would you expect to be from Springfield Middle School?

4. Select a random sample of size 10.

5. Did your random sample have the expected number of scores from Springfield Middle School?

 Activity

20.3 Estimating a Measure of Center for the Population

1. Decide which measure of center makes the most sense to use based on the distribution of your sample. Discuss your thinking with your partner. If you disagree, work to reach an agreement.

2. Estimate this measure of center for your population based on your sample.

3. Calculate the measure of variability for your sample that goes with the measure of center that you found.

 Activity

20.4 Comparing Populations

Using only the values you computed in the previous two activities, compare your sample to your partner's.

Is it reasonable to conclude that the measures of center for each of your populations are meaningfully different? Explain or show your reasoning.

Learning Targets

Lesson	Learning Target(s)
8-1 Mystery Bags	• I can get an idea for the likelihood of an event by using results from previous experiments.
8-2 Chance Experiments	• I can describe the likelihood of events using the words impossible, unlikely, equally likely as not, likely, or certain. • I can tell which event is more likely when the chances of different events are expressed as fractions, decimals, or percentages.
8-3 What are Probabilities?	• I can use the sample space to calculate the probability of an event when all outcomes are equally likely. • I can write out the sample space for a simple chance experiment.

(continued on the next page)

(continued from the previous page)

Lesson	Learning Target(s)
8-4 Estimating Probabilities through Repeated Experiments	• I can estimate the probability of an event based on the results from repeating an experiment. • I can explain whether certain results from repeated experiments would be surprising or not.
8-5 More Estimating Probabilities	• I can calculate the probability of an event when the outcomes in the sample space are not equally likely. • I can explain why results from repeating an experiment may not exactly match the expected probability for an event.
8-6 Estimating Probabilities using Simulation	• I can simulate a real-world situation using a simple experiment that reflects the probability of the actual event.
8-7 Simulating Multi-Step Experiments	• I can use a simulation to estimate the probability of a multi-step event.

Lesson		Learning Target(s)
8-8	Keeping Track of all Possible Outcomes	• I can write out the sample space for a multi-step experiment, using a list, table, or tree diagram.
8-9	Multi-Step Experiments	• I can use the sample space to calculate the probability of an event in a multi-step experiment.
8-10	Designing Simulations	• I can design a simulation to estimate the probability of a multi-step real-world situation.
8-11	Comparing Groups	• I can calculate the difference between two means as a multiple of the mean absolute deviation. • When looking at a pair of dot plots, I can determine whether the distributions are very different or have a lot of overlap.

(continued on the next page)

(continued from the previous page)

Lesson	Learning Target(s)
8-12 Larger Populations	• I can explain why it may be useful to gather data on a sample of a population. • When I read or hear a statistical question, I can name the population of interest and give an example of a sample for that population.
8-13 What Makes a Good Sample?	• I can determine whether a sample is representative of a population by considering the shape, center, and spread of each of them. • I know that some samples may represent the population better than others. • I remember that when a distribution is not symmetric, the median is a better estimate of a typical value than the mean.
8-14 Sampling in a Fair Way	• I can describe ways to get a random sample from a population. • I know that selecting a sample at random is usually a good way to get a representative sample.
8-15 Estimating Population Measures of Center	• I can consider the variability of a sample to get an idea for how accurate my estimate is. • I can estimate the mean or median of a population based on a sample of the population.

Lesson	Learning Target(s)
8-16 Estimating Population Proportions	• I can estimate the proportion of population data that are in a certain category based on a sample.
8-17 More about Sampling Variability	• I can use the means from many samples to judge how accurate an estimate for the population mean is. • I know that as the sample size gets bigger, the sample mean is more likely to be close to the population mean.
8-18 Comparing Populations using Samples	• I can calculate the difference between two medians as a multiple of the interquartile range. • I can determine whether there is a meaningful difference between two populations based on a sample from each population.
8-19 Comparing Populations with Friends	• I can decide what information I need to know to be able to compare two populations based on a sample from each.

(continued on the next page)

(continued from the previous page)

Lesson	Learning Target(s)
8-20 Memory Test	• I can compare two groups by taking a random sample, calculating important measures, and determining whether the populations are meaningfully different.

Notes:

Putting It All Together

Corey Jenkins/Image Source

Let's go for a run! At the end of this unit, you'll apply what you've learned throughout the year to design a 5K running course.

Topics
- Running a Restaurant
- Making Connections
- Designing a Course

Unit 9
Putting It All Together

Lesson 9-1

Planning Recipes

NAME _____ DATE _____ PERIOD _____

Learning Goal Let's choose some recipes for a restaurant.

Activity
1.1 A Recipe for Your Restaurant

Imagine you could open a restaurant.

1. Select a recipe for a main dish you would like to serve at your restaurant.

2. Record the amount of each ingredient from your recipe in the first two columns of the table. You may not need to use every row.

Ingredient	Amount	Amount per Serving	Calories per Serving

3. How many servings does this recipe make? Determine the amount of each ingredient in one serving, and record it in the third column of the table.

4. Restaurants are asked to label how many calories are in each meal on their menu.

   a. Use the nutrition information to calculate the amount of calories from each ingredient in your meal, and record it in the last column of the table.

   b. Next, find the total calories in one serving of your meal.

5. If a person wants to eat 2,000 calories per day, what percentage of their daily calorie intake would one serving of your meal be?

NAME _____ DATE _____ PERIOD _____

Grains

Ingredient	Mass (g)	Calories	Fat (g)	Sodium (mg)
Biscuits, refrigerated dough (1)	58	178	6.14	567
Bread Crumbs (1 oz)	28.35	112	1.5	208
Cornmeal (1 c)	157	581	2.75	11
Egg Noodles (1 c)	38	146	1.69	8
Hamburger or Hotdog Buns (1)	44	123	1.72	217
Oats (1 c)	81	307	5.28	5
Pasta (1 c)	91	338	1.37	5
Pie Crust, refrigerated (1)	229	1019	58.3	937
Pita Bread (1 oz)	28.35	8	0.34	152
Quinoa (1 c)	170	626	10.32	8
Ramen Noodles (1 pkg)	81	356	14.25	1503
Rice Flour (1 c)	158	578	2.24	0
Rice, brown (1 c)	185	679	5.92	9
Rice, white (1 c)	185	675	1.22	9
Saltine Crackers (5)	14.9	62	1.29	140
Taco Shells (1)	12.9	61	2.81	42
Tortillas (1)	49	146	3.71	364
Wheat Bread (1 slice)	29	79	1.31	137
Wheat Flour (1 c)	125	455	1.22	2
White Bread (1 slice)	29	77	0.97	142

Vegetables

Ingredient	Mass (g)	Calories	Fat (g)	Sodium (mg)
Asparagus (1 c)	134	27	0.16	3
Avocados (1 c)	150	240	22	10
Bell Peppers (1 c)	149	46	0.45	6
Broccoli (1 c)	91	31	0.34	30
Carrots (1 c)	128	52	0.31	88
Cauliflower (1 c)	107	27	0.3	32
Celery (1 c)	101	16	0.17	81
Chives (1 tbsp)	3	1	0.02	0
Corn (1 c)	145	125	1.96	22
Cucumber (1 c)	133	16	0.21	3
Green Beans (1 c)	100	31	0.22	6
Lettuce (1 c)	47	8	0.14	7
Mushrooms (1 c)	70	15	0.24	4
Onions (1 c)	160	64	0.16	6
Peas, frozen (1 c)	134	103	0.54	145
Potatoes $\left(\frac{1}{2}\,c\right)$	75	59	0.11	14
Spinach (1 c)	30	7	0.12	24
Squash (1 c)	113	18	0.2	2
Sweet Potatoes (1 c)	133	114	0.07	73
Tomatoes (1 c)	149	27	0.3	7

NAME _____ DATE _____ PERIOD _____

Fruit

Ingredient	Mass (g)	Calories	Fat (g)	Sodium (mg)
Apple Juice (1 c)	248	114	0.32	10
Apples (1 c)	110	53	0.14	0
Bananas (1 c)	225	200	0.74	2
Blueberries (1 c)	148	84	0.49	1
Cantaloupe (1 c)	177	60	0.34	28
Cherries (1 c)	138	87	0.28	0
Cranberries, dried $\left(\frac{1}{4} c\right)$	40	123	0.44	2
Grapes (1 c)	151	104	0.24	3
Lemon Juice (1 c)	244	54	0.59	2
Mandarin Oranges (1 c)	252	154	0.25	15
Mangoes (1 c)	165	99	0.63	2
Orange Juice (1 c)	249	122	0.3	5
Oranges (1 c)	180	85	0.22	0
Peaches (1 c)	154	60	0.38	0
Pears (1 c)	140	80	0.2	1
Pineapple, canned (1 c)	181	109	0.2	2
Pomegranate Juice (1 c)	1249	134	0.72	22
Raisins (1 c)	165	493	0.76	18
Raspberries (1 c)	123	64	0.8	1
Strawberries (1 c)	152	49	0.46	2

Meat

Ingredient	Mass (g)	Calories	Fat (g)	Sodium (mg)
Bacon (1 slice)	26	106	10.21	122
Chicken Thigh (1)	193	427	32.06	156
Chicken, light meat (3 oz)	85	100	1.45	60
Cob (3 oz)	85	61	0.17	93
Crab (3 oz)	85	73	0.82	251
Ground Beef (4 oz)	113	375	33.9	75
Ground Turkey (4 oz)	113	172	9.44	80
Halibut (3 oz)	85	77	1.13	58
Ham (1 oz)	28.35	38	1.53	319
Hot Dogs (1)	51	141	12.33	498
Lobster (1)	150	116	1.12	634
Pepperoni (3 oz)	85	428	39.34	1345
Pork Sausage (1)	25	72	6.2	185
Pork Tenderloin (3 oz)	85	102	3	44
Salmon (1 fillet)	108	373	12.34	55
Shrimp (3 oz)	85	72	0.43	101
Tofu $\left(\frac{1}{2} c\right)$	126	98	5.25	15
Trout (1 fillet)	79	111	4.88	40
Tuna, canned (1 oz)	28.35	24	0.27	70
Turkey (3 oz)	85	92	2.12	105

NAME _____ DATE _____ PERIOD _____

Nuts, Beans, and Seeds

Ingredient	Mass (g)	Calories	Fat (g)	Sodium (mg)
Almonds (1 c)	143	828	71.4	1
Black Beans (1 c)	240	218	0.7	331
Cashews (1 oz)	28.35	157	12.43	3
Chickpeas (1 c)	240	211	4.68	667
Coconut (1 c)	80	283	26.8	16
Fava Beans (1 c)	256	182	0.56	1160
Flaxseed (1 tbsp)	10.3	55	4.34	3
White Beans (1 c)	262	299	1.02	969
Kidney Beans (1 c)	256	215	1.54	758
Lentils (1 c)	192	676	2.04	12
Lima Beans (1 c)	164	216	0.72	85
Macadamia Nuts (1 c)	134	962	101.53	7
Peanut Butter (2 tbsp)	32	191	16.22	136
Peanuts (1 oz)	28.35	166	14.08	116
Pecans (1 c)	109	753	78.45	0
Pinto Beans (1 c)	240	197	1.34	643
Pistachios (1 c)	123	689	55.74	1
Pumpkin Seeds (1 c)	129	721	63.27	9
Sesame Seeds (1 c)	144	825	71.52	16
Sunflower Seeds (1 c)	46	269	23.67	4

Dairy

Ingredient	Mass (g)	Calories	Fat (g)	Sodium (mg)
Almond Milk (1 c)	262	39	2.88	186
Blue Cheese (1 oz)	28.35	100	8.15	325
Butter (1 pat)	5	36	4.06	1
Cheddar Cheese (1 c)	132	533	43.97	862
Coconut Milk (1 c)	226	445	48.21	29
Cream Cheese (1 tbsp)	14.5	51	4.99	46
Egg White (1)	33	17	0.06	55
Egg Yolk (1)	17	55	4.51	8
Eggs (1)	50	72	4.76	71
Evaporated Milk (1 c)	252	270	5.04	252
Whipping Cream (1 c)	120	408	43.3	32
Margarine (1 tbsp)	14.2	101	11.38	4
Milk, skim (1 c)	245	83	0.2	103
Milk, whole (1 c)	244	149	7.93	105
Mozzarella Cheese (1 c)	132	389	26.11	879
Parmesan Cheese (1 c)	100	420	27.84	1804
Sour Cream (1 tbsp)	12	16	1.27	10
Soy Milk (1 c)	243	80	3.91	90
Swiss Cheese (1 c)	132	519	40.91	247
Yogurt (6 oz)	170	107	2.64	119

NAME _____ DATE _____ PERIOD _____

Sauces and Other Liquids

Ingredient	Mass (g)	Calories	Fat (g)	Sodium (mg)
Barbecue Sauce (1 tbsp)	17	29	0.11	175
Chicken Broth (1 c)	249	15	0.52	924
Cream of Chicken Soup $\left(\frac{1}{2} c\right)$	126	113	7.27	885
Gravy (1 c)	233	1	5.5	1305
Honey (1 c)	339	25	0	14
Italian Dressing (1 tbsp)	14.7	35	3.1	146
Jams and Jellies (1 tbsp)	20	56	0.01	6
Ketchup (1 tbsp)	17	17	0.02	154
Mayonnaise (1 tbsp)	15	103	11.67	73
Mustard (1 tsp)	5	3	0.17	55
Pasta Sauce $\left(\frac{1}{2} c\right)$	132	66	2.13	577
Ranch Dressing (1 tbsp)	15	64	6.68	135
Salsa (2 tbsp)	36	10	0.06	256
Soy Sauce (1 tbsp)	16	8	0.09	879
Vanilla Extract (1 tsp)	4.2	12	0	0
Vegetable Broth (1 c)	221	11	0.15	654
Vegetable Oil (1 tbsp)	14	124	14	0
Vinegar (1 tbsp)	15	3	0	0
Water (1 fl oz)	29.6	0	0	1
Worcestershire Sauce (1 tbsp)	17	13	0	167

Spices and Other Powders

Ingredient	Mass (g)	Calories	Fat (g)	Sodium (mg)
Baking Powder (1 tsp)	4.6	2	0	488
Baking Soda (1 tsp)	4.6	0	0	1259
Black Pepper (1 tsp)	2.3	6	0.07	0
Chicken Bouillon (1 cube)	4.8	10	0.23	1152
Chili Powder (1 tsp)	2.7	8	0.39	77
Cinnamon (1 tsp)	2.6	6	0.03	0
Cocoa Powder (1 c)	86	196	11.78	18
Cornstarch (1 c)	128	488	0.06	12
Cumin (1 tsp)	2.1	8	0.47	4
Garlic (1 clove)	3	4	0.01	0.5
Garlic Powder (1 tsp)	3.1	10	0.02	2
Onion Powder (1 tsp)	2.4	8	0.02	2
Onion Soup Mix (1 tbsp)	7.5	22	0.03	602
Oregano (1 tsp)	1	3	0.04	0
Paprika (1 tsp)	2.3	6	0.3	2
Parsley (1 tsp)	0.5	1	0.03	2
Powdered Sugar (1 c)	120	467	0	2
Salt (1 tsp)	6	0	0	2325
Sugar (1 tsp)	2.8	11	0	0
Taco Seasoning (2 tsp)	5.7	18	0	411

NAME _____ DATE _____ PERIOD _____

Are you ready for more?

The labels on packaged foods tell how much of different nutrients they contain. Here is what some different food labels say about their sodium content.

- cheese crackers, 351 mg, 14% daily value

- apple chips, 15 mg, <1% daily value

- granola bar, 82 mg, 3% daily value

Estimate the maximum recommended amount of sodium intake per day (100% daily value). Explain your reasoning.

 **Activity**

1.2 Health Claims

For a meal to be considered:

- "low calorie"— it must have 120 calories or less per 100 grams of food.

- "low fat"— it must have 3 grams of fat or less per 100 grams of food.

- "low sodium"— it must have 140 milligrams of sodium or less per 100 grams of food.

1. Does the meal you chose in the previous activity meet the requirements to be considered:

   a. low calorie?

   b. low fat?

   c. low sodium?

Be prepared to explain your reasoning.

2. Select or invent another recipe you would like to serve at your restaurant that does meet the requirements to be considered either low calorie, low fat, or low sodium. Show that your recipe meets that requirement. Organize your thinking so it can be followed by others.

Ingredient	Amount per Serving	Calories per Serving	Fat per Serving	Sodium per Serving

Lesson 9-2

Costs of Running a Restaurant

NAME _____ DATE _____ PERIOD _____

Learning Goal Let's explore how much the food will cost.

 ## Activity
2.1 Introducing Spreadsheets

1. Type each formula into the cells of a spreadsheet program and press enter. Record what the cell displays. Make sure to type each formula exactly as it is written here.

	A	B	C	D
1	=40−32	=1.5+3.6	=14/7	=0.5*6

2. **a.** Predict what will happen if you type the formula =A1*C1 into cell C2 of your spreadsheet.

   b. Type in the formula, and press enter to check your prediction.

3. **a.** Predict what will happen next if you delete the formula in cell A1 and replace it with the number 100.

   b. Replace the formula with the number, and press enter to check your prediction.

4. **a.** Predict what will happen if you copy cell C2 and paste it into cell D2 of your spreadsheet.

   b. Copy and paste the formula to check your prediction.

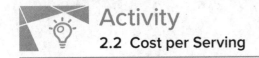

Activity

2.2 Cost per Serving

1. Set up a spreadsheet with these column labels in the first row.

	A	B	C	D
1	Ingredient	Unit in Recipe	Amount in Recipe	Amount per Serving
2				

    a. Type the information about the ingredients in your recipe into the first 3 columns of the spreadsheet.

    b. Type a formula into cell D2 to automatically calculate the amount per serving for your first ingredient.

    c. Copy cell D2 and paste it into the cells beneath it to calculate the amount per serving for the rest of your ingredients. Pause here so your teacher can review your work.

NAME _____ DATE _____ PERIOD _____

2. Add these column labels to your spreadsheet.

E	F	G	H
Purchase Price	Purchase Amount	Purchase Unit	Cost per Purchase Unit

a. Research the cost of each ingredient in your meal, and record the information in columns E, F, and G.

b. Type a formula into cell H2 to automatically calculate the cost per purchase unit for your first ingredient.

c. Copy cell H2, and paste it into the cells beneath it to calculate the cost per purchase unit for the rest of your ingredients.

3. Add these column labels to your spreadsheet.

I	J	K
Conversion from Purchase Units to Recipe Units	Cost per Recipe Unit	Cost per Serving

a. Complete column I with how many of your recipe unit are in 1 of your purchase unit for each ingredient. For example, if your recipe unit was cups and your purchase unit was gallons, then your conversion would be 16 because there are 16 cups in 1 gallon.

b. Type a formula into cell J2 to calculate the cost per recipe unit for your first ingredient.

c. Type a formula into cell K2 to calculate the cost per serving for your first ingredient.

d. Compare formulas with your partner. Discuss your thinking. If you disagree, work to reach an agreement.

e. Copy cells J2 and K2, and paste them into the cells beneath them to calculate the cost per recipe unit and cost per serving for the rest of your ingredients.

4. Type a formula into the first empty cell below your last ingredient in column K to calculate the total cost per serving for all of the ingredients in your recipe. Record the answer here.

Lesson 9-3

More Costs of Running a Restaurant

NAME _____ DATE _____ PERIOD _____

Learning Goal Let's explore how much it costs to run a restaurant.

 ## Activity

3.1 Are We Making Money?

1. Restaurants have many more expenses than just the cost of the food.

   a. Make a list of other items you would have to spend money on if you were running a restaurant.

   b. Identify which expenses on your list depend on the number of meals ordered and which are independent of the number of meals ordered.

   c. Identify which of the expenses that are independent of the number of meals ordered only have to be paid once and which are ongoing.

   d. Estimate the monthly cost for each of the ongoing expenses on your list. Next, calculate the total of these monthly expenses.

2. Tell whether each restaurant is making a profit or losing money if they have to pay the amount you predicted in ongoing expenses per month. Organize your thinking so it can be followed by others.

   a. Restaurant A sells 6,000 meals in one month, at an average price of $17 per meal and an average cost of $4.60 per meal.

   b. Restaurant B sells 8,500 meals in one month, at an average price of $8 per meal and an average cost of $2.20 per meal.

   c. Restaurant C sells 4,800 meals in one month, at an average price of $29 per meal and an average cost of $6.90 per meal.

3. Respond to each of the following.

   a. Predict how many meals your restaurant would sell in one month.

   b. How much money would you need to charge for each meal to be able to cover all the ongoing costs of running a restaurant?

4. What percentage of the cost of the ingredients is the markup on your meal?

NAME _____ DATE _____ PERIOD _____

Activity

3.2 Disposable or Reusable?

A sample of full service restaurants and a sample of fast food restaurants were surveyed about the average number of customers they serve per day.

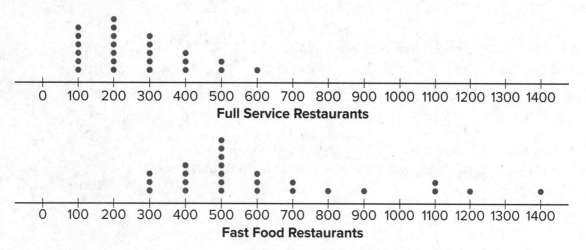

1. How does the average number of customers served per day at a full service restaurant generally compare to the number served at a fast food restaurant? Explain your reasoning.

2. About how many customers do you think your restaurant will serve per day? Explain your reasoning.

3. Here are prices for plates and forks.

   a. Using your predicted number of customers per day from the previous question, write an equation for the total cost, d, of using disposable plates and forks for every customer for n days.

	Plates	Forks
Disposable	165 paper plates for $12.50	600 plastic forks for $10
Reusable	12 ceramic plates for $28.80	24 metal forks for $30

   b. Is d proportional to n? Explain your reasoning.

   c. Use your equation to predict the cost of using disposable plates and forks for 1 year. Explain any assumptions you make with this calculation.

4. Respond to each of the following.

   a. How much would it cost to buy enough reusable plates and forks for your predicted number of customers per day?

   b. If it costs $10.75 a day to wash the reusable plates and forks, write an expression that represents the total cost, r, of buying and washing reusable plates and forks after n days.

   c. Is r proportional to n? Explain your reasoning.

   d. How many days can you use the reusable plates and forks for the same cost that you calculated for using disposable plates and forks for 1 year?

Lesson 9-4

Restaurant Floor Plan

NAME _____ DATE _____ PERIOD _____

Learning Goal Let's design the floor plan for a restaurant.

Activity

4.1 Dining Area

1. Restaurant owners say it is good for each customer to have about 300 in²
   of space at their table. How many customers would you seat at
   each table?

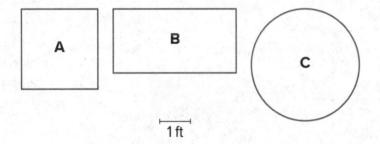

1 ft

2. It is good to have about 15 ft² of floor space per customer in the
   dining area.

   a. How many customers would you like to be able to seat at one time?

   b. What size and shape dining area would be large enough to fit
      that many customers?

c. Select an appropriate scale, and create a scale drawing of the outline of your dining area.

3. Using the same scale, what size would each of the tables from the first question appear on your scale drawing?

NAME _____ DATE _____ PERIOD _____

4. To ensure fast service, it is good for all of the tables to be within 60 ft of the place where the servers bring the food out of the kitchen. Decide where the food pickup area will be, and draw it on your scale drawing. Next, show the limit of how far away tables can be positioned from this place.

5. It is good to have at least $1\frac{1}{2}$ ft between each table and at least $3\frac{1}{2}$ ft between the sides of tables where the customers will be sitting. On your scale drawing, show one way you could arrange tables in your dining area.

Are you ready for more?

The dining area usually takes up about 60% of the overall space of a restaurant because there also needs to be room for the kitchen, storage areas, office, and bathrooms. Given the size of your dining area, how much more space would you need for these other areas?

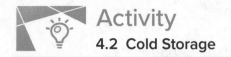

Activity

4.2 Cold Storage

Some restaurants have very large refrigerators or freezers that are like small rooms. The energy to keep these rooms cold can be expensive.

- A standard walk-in refrigerator (rectangular, 10 feet wide, 10 feet long, and 7 feet tall) will cost about $150 per month to keep cold.

- A standard walk-in freezer (rectangular, 8 feet wide, 10 feet long, and 7 feet tall) will cost about $372 per month to keep cold.

Here is a scale drawing of a walk-in refrigerator and freezer. About how much would it cost to keep them both cold? Show your reasoning.

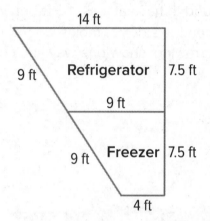

Lesson 9-5

How Crowded is This Neighborhood?

NAME _____ DATE _____ PERIOD _____

Learning Goal Let's see how proportional relationships apply to where people live.

Activity

5.1 Dot Density

The figure shows four squares. Each square encloses an array of dots. Squares A and B have side length 2 inches. Squares C and D have side length 1 inch.

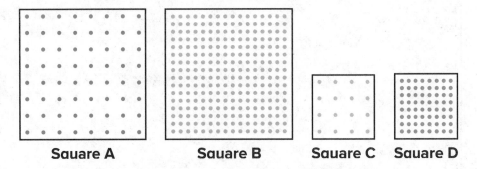

Square A   Square B   Square C   Square D

1. Complete the table with information about each square.

Square	Area of the Square in Square Inches	Number of Dots	Number of Dots per Square Inch
A			
B			
C			
D			

2. Compare each square to the others. What is the same and what is different?

The figure shows two arrays, each enclosed by a square that is 2 inches wide.

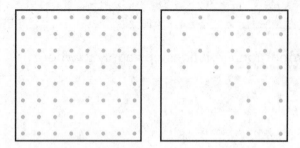

1. Let a be the area of the square and d be the number of dots enclosed by the square. For each square, plot a point that represents its values of a and d.

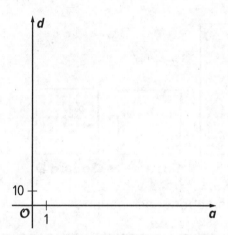

2. Draw lines from (0,0) to each point. For each line, write an equation that represents the proportional relationship.

3. What is the constant of proportionality for each relationship? What do the constants of proportionality tell us about the dots and squares?

NAME _____ DATE _____ PERIOD _____

Activity
5.3 Housing Density

Here are pictures of two different neighborhoods.

This image depicts an area that is 0.3 kilometers long and 0.2 kilometers wide.

0.1 km

This image depicts an area that is 0.4 kilometers long and 0.2 kilometers wide.

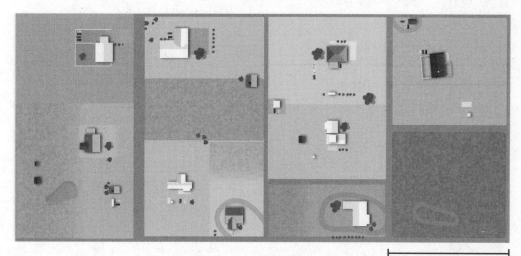

0.1 km

For each neighborhood, find the number of houses per square kilometer.

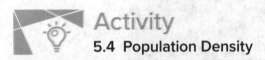

Activity

5.4 Population Density

- New York City has a population of 8,406 thousand people and covers an area of 1,214 square kilometers.

- Los Angeles has a population of 3,884 thousand people and covers an area of 1,302 square kilometers.

1. The points labeled *A* and *B* each correspond to one of the two cities. Which is which? Label them on the graph.

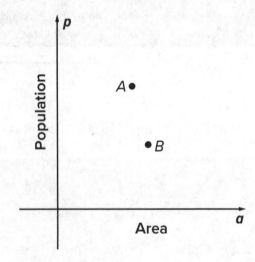

NAME _____ DATE _____ PERIOD _____

2. Write an equation for the line that passes through (0,0) and *A*. What is the constant of proportionality?

3. Write an equation for the line that passes through (0,0) and *B*. What is the constant of proportionality?

4. What do the constants of proportionality tell you about the crowdedness of these two cities?

1. Predict where these types of regions would be shown on the graph:

   a. a suburban region where houses are far apart, with big yards

   b. a neighborhood in an urban area with many high-rise apartment buildings

   c. a rural state with lots of open land and not many people

2. Next, use this data to check your predictions:

Place	Description	Population	Area (km²)
Chalco	a suburb of Omaha, Nebraska	10,994	7.5
Anoka County	a county in Minnesota, near Minneapolis/St. Paul	339,534	1,155
Guttenberg	a city in New Jersey	11,176	0.49
New York	a state	19,746,227	141,300
Rhode Island	a state	1,055,173	3,140
Alaska	a state	736,732	1,717,856
Tok	a community in Alaska	1,258	342.7

Lesson 9-6
Fermi Problems

NAME _____ DATE _____ PERIOD _____

Learning Goal Let's estimate some quantities.

 **Activity**
6.1 How Old Are You?

What is your *exact* age at this moment?

 **Activity**
6.2 A Heart Stoppingly Large Number

How many times has your heart beat in your lifetime?

Activity

6.3 All the Hairs on Your Head

How many strands of hair do you have on your head?

Lesson 9-7

More Expressions and Equations

NAME _____ DATE _____ PERIOD _____

Learning Goal Let's solve harder problems by writing equivalent expressions.

Activity

7.1 Tickets for the School Play

Student tickets for the school play cost $2 less than adult tickets.

1. If a represents the price of one adult ticket, write an expression for the price of a student ticket.

2. Write an expression that represents the amount of money they collected each night.

   a. The first night, the school sold 60 adult tickets and 94 student tickets.

   b. The second night, the school sold 83 adult tickets and 127 student tickets.

3. Write an expression that represents the total amount of money collected from ticket sales on both nights.

4. Over these two nights, they collected a total of $1,651 in ticket sales.

   a. Write an equation that represents this situation.

   b. What was the cost of each type of ticket?

5. Is your solution reasonable? Explain how you know.

 ## Activity

7.2 A Souvenir Stand

The souvenir stand sells hats, postcards, and magnets. They have twice as many postcards as hats, and 100 more magnets than post cards.

1. Let h represent the total number of hats. Write an expression in terms of h for the total number of items they have to sell.

2. The owner of the stand pays $8 for each hat, $0.10 for each post card, and $0.50 for each magnet. Write an expression for the total cost of the items.

NAME _____ DATE _____ PERIOD _____

3. The souvenir stand sells the hats for $11.75 each, the postcards for $0.25 each, and the magnets for $3.50 each. Write an expression for the total amount of money they would take in if they sold all the items.

4. Profits are calculated by subtracting costs from income. Write an expression for the profits of the souvenir stand if they sell all the items they have. Use properties to write an equivalent expression with fewer terms.

5. The souvenir stand sells all these items and makes a total profit of $953.25.

   a. Write an equation that represents this situation.

   b. How many of each item does the souvenir stand sell? Explain or show your reasoning.

Activity

7.3 Jada Crochets a Scarf

Basic crochet stitches are called single, double, and triple. Jada measures her average stitch size and sees that a "double crochet" stitch is not really twice as long; it uses $\frac{1}{2}$ inch less than twice as much yarn as a single crochet stitch. Jada's "triple crochet" stitch uses 1 inch less than three times as much yarn as a single crochet stitch.

1. Write an expression that represents the amount of yarn Jada needs to crochet a scarf that includes 800 single crochet stitches, 400 double crochet stitches, and 200 triple crochet stitches.

2. Write an equivalent expression with as few terms as possible.

3. If Jada uses 5,540 inches of yarn for the entire scarf, what length of yarn does she use for a single crochet stitch?

Lesson 9-8

Measurement Error (Part 1)

NAME _____ DATE _____ PERIOD _____

Learning Goal Let's check how accurate our measurements are.

 Activity

8.1 How Long Are These Pencils?

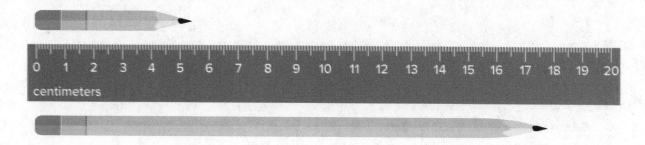

1. Estimate the length of each pencil.

2. How accurate are your estimates?

3. For each estimate, what is the largest possible percent error?

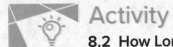

Activity

8.2 How Long Are These Floor Boards?

A wood floor is made by laying multiple boards end to end. Each board is measured with a maximum percent error of 5%. What is the maximum percent error for the total length of the floor?

Lesson 9-9

Measurement Error (Part 2)

NAME _____ DATE _____ PERIOD _____

Learning Goal Let's check how accurate our calculations are.

 ## Activity

9.1 Measurement Error for Area

Imagine that you measure the length and width of a rectangle and you know the measurements are accurate within 5% of the actual measurements. If you use your measurements to find the area, what is the maximum percent error for the area of the rectangle?

Activity

9.2 Measurement Error for Volume

1. The length, width, and height of a rectangular prism were measured to be 10 cm, 12 cm, and 25 cm. Assuming that these measurements are accurate to the nearest cm, what is the largest percent error possible for:

   a. each of the dimensions?

   b. the volume of the prism?

2. If the length, width, and height of a right rectangular prism have a maximum percent error of 1%, what is the largest percent error possible for the volume of the prism?

Lesson 9-10

Measuring Long Distances over Uneven Terrain

NAME _____ DATE _____ PERIOD _____

Learning Goal Let's measure long distances over uneven terrain.

Activity
10.1 How Far Is It?

How do people measure distances in different situations? What tools do they use? Come up with at least three different methods and situations where those methods are used.

Activity
10.2 Planning a 5K Course

The school is considering holding a 5K fundraising walk on the school grounds. Your class is supposed to design the course for the walk.

1. What will you need to do to design the course for the walk?

2. Come up with a method to measure the course. Pause here so your teacher can review your plan.

Activity

10.3 Comparing Methods

Let's see how close different measuring methods are to each other. Your teacher will show you a path to measure.

1. Use your method to measure the length of the path at least two times.

2. Decide what distance you will report to the class.

3. Compare your results with those of two other groups. Express the differences between the measurements in terms of percentages.

4. Discuss the advantages and disadvantages of each group's method.

Lesson 9-11

Building a Trundle Wheel

NAME _____ DATE _____ PERIOD _____

Learning Goal Let's build a trundle wheel.

Activity

11.1 What Is a Trundle Wheel?

A tool that surveyors use to measure distances is called a trundle wheel.

1. How does a trundle wheel measure distance?

2. Why is this method of measuring distances better than the methods we used in the previous lesson?

3. How could we construct a simple trundle wheel? What materials would we need?

Activity

11.2 Building a Trundle Wheel

Your teacher will give you some supplies. Construct a trundle wheel and use it to measure the length of the classroom. Record:

1. the diameter of your trundle wheel

2. the number of clicks across the classroom

3. the length of the classroom (Be prepared to explain your reasoning.)

Lesson 9-12

Using a Trundle Wheel to Measure Distances

NAME _____ DATE _____ PERIOD _____

Learning Goal Let's use our trundle wheels.

 ## Activity

12.1 Measuring Distances with the Trundle Wheel

Earlier you made trundle wheels so that you can measure long distances. Your teacher will show you a path to measure.

1. Measure the path with your trundle wheel three times and calculate the distance. Record your results in the table.

Trial Number	Number of Clicks	Computation	Distance
1			
2			
3			

2. Decide what distance you will report to the class. Be prepared to explain your reasoning.

3. Compare this distance with the distance you measured the other day for this same path.

4. Compare your results with the results of two other groups. Express the differences between the measurements in terms of percentages.

Lesson 9-13

Designing a 5K Course

NAME _____ DATE _____ PERIOD _____

Learning Goal Let's map out the 5K course.

Activity

13.1 Make a Proposal

Your teacher will give you a map of the school grounds.

1. On the map, draw in the path you measured earlier with your trundle wheel and label its length.

2. Invent another route for a walking course and draw it on your map. Estimate the length of the course you drew.

3. How many laps around your course must someone complete to walk 5 km?

Activity

13.2 Measuring and Finalizing the Course

1. Measure your proposed race course with your trundle wheel at least two times. Decide what distance you will report to the class.

2. Revise your course, if needed.

3. Create a visual display that includes:
   - a map of your final course
   - the starting and ending locations
   - the number of laps needed to walk 5 km
   - any other information you think would be helpful to the race organizers

Are you ready for more?

The map your teacher gave you didn't include a scale. Create one.

Glossary

A

absolute value The absolute value of a number is its distance from 0 on the number line.

The absolute value of -7 is 7, because it is 7 units away from 0. The absolute value of 5 is 5, because it is 5 units away from 0.

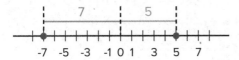

adjacent angles Adjacent angles share a side and a vertex.

In this diagram, angle *ABC* is adjacent to angle *DBC*.

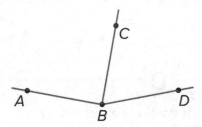

area Area is the number of square units that cover a two-dimensional region, without any gaps or overlaps.

For example, the area of region A is 8 square units. The area of the shaded region of B is $\frac{1}{2}$ square unit.

Region A

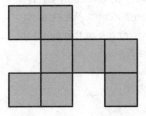

Region B

area of a circle If the radius of a circle is *r* units, then the area of the circle is πr^2 square units.

For example, a circle has radius 3 inches. Its area is $\pi 3^2$ square inches, or 9π square inches, which is approximately 28.3 square inches.

B

base (of a prism or pyramid) The word *base* can also refer to a face of a polyhedron. A prism has two identical bases that are parallel. A pyramid has one base. A prism or pyramid is named for the shape of its base.

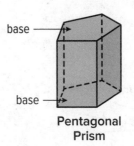

Pentagonal Prism

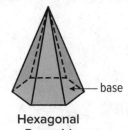

Hexagonal Pyramid

C

chance experiment A chance experiment is something you can do over and over again, and you don't know what will happen each time.

For example, each time you spin the spinner, it could land on red, yellow, blue, or green.

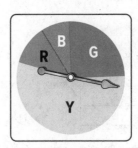

circle A circle is made out of all the points that are the same distance from a given point.

For example, every point on this circle is 5 cm away from point *A*, which is the center of the circle.

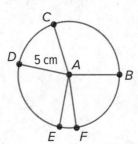

circumference The circumference of a circle is the distance around the circle. If you imagine the circle as a piece of string, it is the length of the string. If the circle has radius r then the circumference is $2\pi r$.

The circumference of a circle of radius 3 is $2 \cdot \pi \cdot 3$, which is 6π, or about 18.85.

complementary Complementary angles have measures that add up to 90 degrees.

For example, a 15° angle and a 75° angle are complementary.

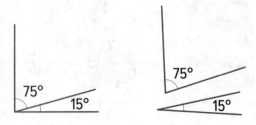

constant of proportionality In a proportional relationship, the values for one quantity are each multiplied by the same number to get the values for the other quantity. This number is called the constant of proportionality.

In this example, the constant of proportionality is 3, because $2 \cdot 3 = 6$, $3 \cdot 3 = 9$, and $5 \cdot 3 = 15$. This means that there are 3 apples for every 1 orange in the fruit salad.

Number of Oranges	Number of Apples
2	6
3	9
5	15

coordinate plane The coordinate plane is a system for telling where points are. For example, point R is located at (3, 2) on the coordinate plane, because it is three units to the right and two units up.

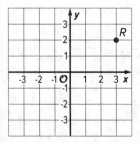

corresponding When part of an original figure matches up with part of a copy, we call them corresponding parts. These could be points, segments, angles, or distances.

For example, point B in the first triangle corresponds to point E in the second triangle. Segment AC corresponds to segment DF.

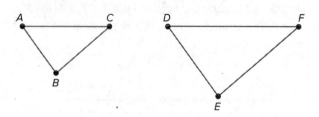

cross section A cross section is the new face you see when you slice through a three-dimensional figure.

For example, if you slice a rectangular pyramid parallel to the base, you get a smaller rectangle as the cross section.

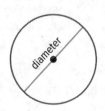

D

deposit When you put money into an account, it is called a *deposit*.

For example, a person added $60 to their bank account. Before the deposit, they had $435. After the deposit, they had $495, because $435 + 60 = 495$.

diameter A diameter is a line segment that goes from one edge of a circle to the other and passes through the center. A diameter can go in any direction. Every diameter of the circle is the same length. We also use the word *diameter* to mean the length of this segment.

E

equivalent expressions Equivalent expressions are always equal to each other. If the expressions have variables, they are equal whenever the same value is used for the variable in each expression.

For example, $3x + 4x$ is equivalent to $5x + 2x$. No matter what value we use for x, these expressions are always equal. When x is 3, both expressions equal 21. When x is 10, both expressions equal 70.

equivalent ratios Two ratios are equivalent if you can multiply each of the numbers in the first ratio by the same factor to get the numbers in the second ratio. For example, $8 : 6$ is equivalent to $4 : 3$, because $8 \cdot \frac{1}{2} = 4$ and $6 \cdot \frac{1}{2} = 3$.

A recipe for lemonade says to use 8 cups of water and 6 lemons. If we use 4 cups of water and 3 lemons, it will make half as much lemonade. Both recipes taste the same, because $8 : 6$ and $4 : 3$ are equivalent ratios.

Cups of Water	Number of Lemons
8	6
4	3

event An event is a set of one or more outcomes in a chance experiment. For example, if we roll a number cube, there are six possible outcomes.

Examples of events are "rolling a number less than 3," "rolling an even number," or "rolling a 5."

expand To expand an expression, we use the distributive property to rewrite a product as a sum. The new expression is equivalent to the original expression.

For example, we can expand the expression $5(4x + 7)$ to get the equivalent expression $20x + 35$.

F

factor (an expression) To factor an expression, we use the distributive property to rewrite a sum as a product. The new expression is equivalent to the original expression.

For example, we can factor the expression $20x + 35$ to get the equivalent expression $5(4x + 7)$.

I

interquartile range (IQR) The interquartile range is one way to measure how spread out a data set is. We sometimes call this the IQR. To find the interquartile range we subtract the first quartile from the third quartile.

For example, the IQR of this data set is 20 because $50 - 30 = 20$.

22	29	30	31	32	43	44	45	50	50	59
		Q1			Q2			Q3		

L

long division Long division is a way to show the steps for dividing numbers in decimal form. It finds the quotient one digit at a time, from left to right.

For example, here is the long division for $57 \div 4$.

$$
\begin{array}{r}
14.25 \\
4\overline{)57.00} \\
-4 \\
\hline
17 \\
-16 \\
\hline
10 \\
-8 \\
\hline
20 \\
-20 \\
\hline
0
\end{array}
$$

M

mean The mean is one way to measure the center of a data set. We can think of it as a balance point. For example, for the data set 7, 9, 12, 13, 14, the mean is 11.

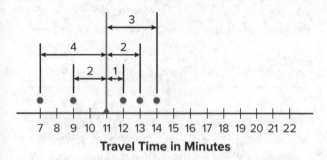

Travel Time in Minutes

To find the mean, add up all the numbers in the data set. Then, divide by how many numbers there are. $7 + 9 + 12 + 13 + 14 = 55$ and $55 \div 5 = 11$.

mean absolute deviation (MAD) The mean absolute deviation is one way to measure how spread out a data set is. Sometimes we call this the MAD. For example, for the data set 7, 9, 12, 13, 14, the MAD is 2.4. This tells us that these travel times are typically 2.4 minutes away from the mean, which is 11.

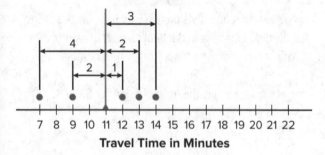

Travel Time in Minutes

To find the MAD, add up the distance between each data point and the mean. Then, divide by how many numbers there are. $4 + 2 + 1 + 2 + 3 = 12$ and $12 \div 5 = 2.4$.

measurement error Measurement error is the positive difference between a measured amount and the actual amount.

For example, Diego measures a line segment and gets 5.3 cm. The actual length of the segment is really 5.32 cm. The measurement error is 0.02 cm, because $5.32 - 5.3 = 0.02$.

median The median is one way to measure the center of a data set. It is the middle number when the data set is listed in order.

For the data set 7, 9, 12, 13, 14, the median is 12.

For the data set 3, 5, 6, 8, 11, 12, there are two numbers in the middle. The median is the average of these two numbers. $6 + 8 = 14$ and $14 \div 2 = 7$.

N

negative number A negative number is a number that is less than zero. On a horizontal number line, negative numbers are usually shown to the left of 0.

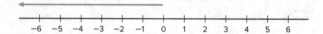

O

origin The origin is the point (0, 0) in the coordinate plane. This is where the horizontal axis and the vertical axis cross.

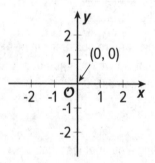

outcome An outcome of a chance experiment is one of the things that can happen when you do the experiment. For example, the possible outcomes of tossing a coin are heads and tails.

P

percent error Percent error is a way to describe error, expressed as a percentage of the actual amount.

For example, a box is supposed to have 150 folders in it. Clare counts only 147 folders in the box. This is an error of 3 folders. The percent error is 2%, because 3 is 2% of 150.

percentage A percentage is a rate per 100.

For example, a fish tank can hold 36 liters. Right now there is 27 liters of water in the tank. The percentage of the tank that is full is 75%.

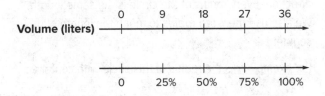

percentage decrease A percentage decrease tells how much a quantity went down, expressed as a percentage of the starting amount.

For example, a store had 64 hats in stock on Friday. They had 48 hats left on Saturday. The amount went down by 16.

This was a 25% decrease, because 16 is 25% of 64.

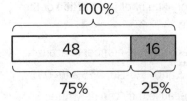

percentage increase A percentage increase tells how much a quantity went up, expressed as a percentage of the starting amount.

For example, Elena had $50 in the bank on Monday. She had $56 on Tuesday. The amount went up by $6.

This was a 12% increase, because 6 is 12% of 50.

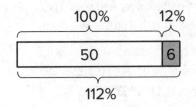

pi (π) There is a proportional relationship between the diameter and circumference of any circle. The constant of proportionality is pi. The symbol for pi is π.

We can represent this relationship with the equation $C = \pi d$, where C represents the circumference and d represents the diameter.

Some approximations for π are $\frac{22}{7}$, 3.14, and 3.14159.

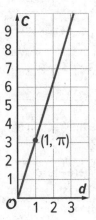

population A population is a set of people or things that we want to study.

For example, if we want to study the heights of people on different sports teams, the population would be all the people on the teams.

positive number A positive number is a number that is greater than zero. On a horizontal number line, positive numbers are usually shown to the right of 0.

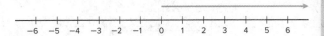

prism A prism is a type of polyhedron that has two bases that are identical copies of each other. The bases are connected by rectangles or parallelograms.

Here are some drawings of prisms.

 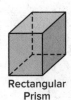

Triangular Prism   Pentagonal Prism   Rectangular Prism

probability The probability of an event is a number that tells how likely it is to happen. A probability of 1 means the event will always happen. A probability of 0 means the event will never happen.

For example, the probability of selecting a moon block at random from this bag is $\frac{4}{5}$.

proportion A proportion of a data set is the fraction of the data in a given category.

For example, a class has 20 students. There are 2 left-handed students and 18 right-handed students in the class. The proportion of students who are left-handed is $\frac{2}{20}$, or 0.1.

proportional relationship In a proportional relationship, the values for one quantity are each multiplied by the same number to get the values for the other quantity.

For example, in this table every value of p is equal to 4 times the value of s on the same row. We can write this relationship as $p = 4s$. This equation shows that s is proportional to p.

s	p
2	8
3	12
5	20
10	40

pyramid A pyramid is a type of polyhedron that has one base. All the other faces are triangles, and they all meet at a single vertex.

Here are some drawings of pyramids.

Rectangular Pyramid

Hexagonal Pyramid

Heptagonal Pyramid

R

radius A radius is a line segment that goes from the center to the edge of a circle. A radius can go in any direction. Every radius of the circle is the same length. We also use the word *radius* to mean the length of this segment.

For example, r is the radius of this circle with center O.

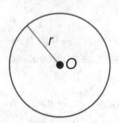

random Outcomes of a chance experiment are random if they are all equally likely to happen.

rational number A rational number is a fraction or the opposite of a fraction.

For example, 8 and -8 are rational numbers because they can be written as $\frac{8}{1}$ and $-\frac{8}{1}$. Also, 0.75 and -0.75 are rational numbers because they can be written as $\frac{75}{100}$ and $-\frac{75}{100}$.

reciprocal Dividing 1 by a number gives the reciprocal of that number. For example, the reciprocal of 12 is $\frac{1}{12}$, and the reciprocal of $\frac{2}{5}$ is $\frac{5}{2}$.

repeating decimal A repeating decimal has digits that keep going in the same pattern over and over. The repeating digits are marked with a line above them.

For example, the decimal representation for $\frac{1}{3}$ is $0.\overline{3}$, which means 0.3333333...The decimal representation for $\frac{25}{22}$ is $1.1\overline{36}$ which means 1.136363636...

representative A sample is representative of a population if its distribution resembles the population's distribution in center, shape, and spread.

For example, this dot plot represents a population.

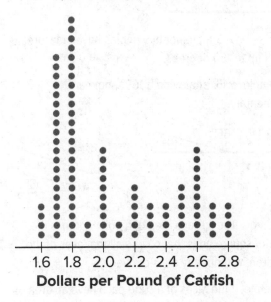

Dollars per Pound of Catfish

This dot plot shows a sample that is representative of the population.

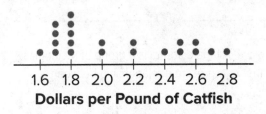

Dollars per Pound of Catfish

right angle A right angle is half of a straight angle. It measures 90 degrees.

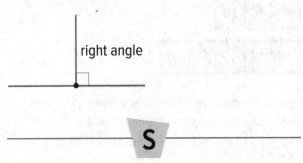

right angle

sample A sample is part of a population. For example, a population could be all the seventh grade students at one school. One sample of that population is all the seventh grade students who are in band.

sample space The sample space is the list of every possible outcome for a chance experiment.

For example, the sample space for tossing two coins is:

heads-heads	tails-heads
heads-tails	tails-tails

scale A scale tells how the measurements in a scale drawing represent the actual measurements of the object.

For example, the scale on this floor plan tells us that 1 inch on the drawing represents 8 feet in the actual room. This means that 2 inches would represent 16 feet, and $\frac{1}{2}$ inch would represent 4 feet.

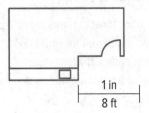

1 in
8 ft

scale drawing A scale drawing represents an actual place or object. All the measurements in the drawing correspond to the measurements of the actual object by the same scale.

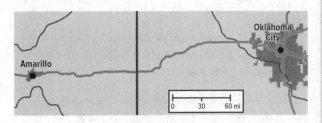

scale factor To create a scaled copy, we multiply all the lengths in the original figure by the same number. This number is called the scale factor.

In this example, the scale factor is 1.5, because $4 \cdot (1.5) = 6$, $5 \cdot (1.5) = 7.5$, and $6 \cdot (1.5) = 9$.

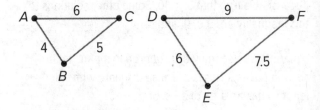

scaled copy A scaled copy is a copy of a figure where every length in the original figure is multiplied by the same number.

For example, triangle *DEF* is a scaled copy of triangle *ABC*. Each side length on triangle *ABC* was multiplied by 1.5 to get the corresponding side length on triangle *DEF*.

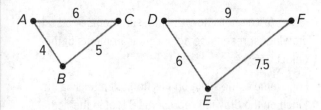

simulation A simulation is an experiment that is used to estimate the probability of a real-world event.

For example, suppose the weather forecast says there is a 25% chance of rain. We can simulate this situation with a spinner with four equal sections. If the spinner stops on red, it represents rain. If the spinner stops on any other color, it represents no rain.

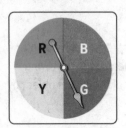

solution to an equation A solution to an equation is a number that can be used in place of the variable to make the equation true.

For example, 7 is the solution to the equation $m + 1 = 8$, because it is true that $7 + 1 = 8$. The solution to $m + 1 = 8$ is not 9, because $9 + 1 \neq 8$.

solution to an inequality A solution to an inequality is a number that can be used in place of the variable to make the inequality true.

For example, 5 is a solution to the inequality $c < 10$, because it is true that $5 < 10$. Some other solutions to this inequality are 9.9, 0, and -4.

squared We use the word *squared* to mean "to the second power." This is because a square with side length s has an area of $s \cdot s$, or s^2.

straight angle A straight angle is an angle that forms a straight line. It measures 180 degrees.

straight angle

supplementary Supplementary angles have measures that add up to 180 degrees.

For example, a 15° angle and a 165° angle are supplementary.

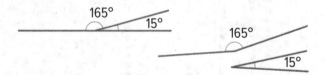

surface area The surface area of a polyhedron is the number of square units that covers all the faces of the polyhedron, without any gaps or overlaps.

For example, if the faces of a cube each have an area of 9 cm^2, then the surface area of the cube is 6 • 9, or 54 cm^2.

tape diagram A tape diagram is a group of rectangles put together to represent a relationship between quantities.

For example, this tape diagram shows a ratio of 30 gallons of yellow paint to 50 gallons of blue paint.

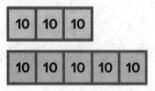

If each rectangle were labeled 5, instead of 10, then the same picture could represent the equivalent ratio of 15 gallons of yellow paint to 25 gallons of blue paint.

term A term is a part of an expression. It can be a single number, a variable, or a number and a variable that are multiplied together. For example, the expression $5x + 18$ has two terms. The first term is $5x$ and the second term is 18.

unit rate A unit rate is a rate per 1.

For example, 12 people share 2 pies equally. One unit rate is 6 people per pie, because $12 \div 2 = 6$. The other unit rate is $\frac{1}{6}$ of a pie per person, because $2 \div 12 = \frac{1}{6}$.

variable A variable is a letter that represents a number. You can choose different numbers for the value of the variable.

For example, in the expression $10 - x$, the variable is x. If the value of x is 3, then $10 - x = 7$, because $10 - 3 = 7$. If the value of x is 6, then $10 - x = 4$, because $10 - 6 = 4$.

vertical angles Vertical angles are opposite angles that share the same vertex. They are formed by a pair of intersecting lines. Their angle measures are equal.

For example, angles *AEC* and *DEB* are vertical angles. If angle *AEC* measures 120°, then angle *DEB* must also measure 120°.

Angles *AED* and *BEC* are another pair of vertical angles.

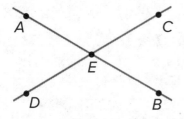

volume Volume is the number of cubic units that fill a three-dimensional region, without any gaps or overlaps.

For example, the volume of this rectangular prism is 60 units³, because it is composed of 3 layers that are each 20 units³.

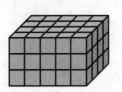

withdrawal When you take money out of an account, it is called a *withdrawal*.

For example, a person removed $25 from their bank account. Before the withdrawal, they had $350. After the withdrawal, they had $325, because $350 - 25 = 325$.

Index

A

absolute value, 353

additive inverse, *See opposites.*

adjacent angles, 614

angles
    adjacent, 614
    complementary, 622
    corresponding, 12–13
    right, 614
    straight, 614
    supplementary, 622
    vertical, 628

area, 40
    distinguishing from
      circumference, 241
    of circles, 222
    relating to circumference, 228
    scale drawings and, 40

axis (axes), 149

B

balanced hangers, *See hanger diagrams.*

base
    of a prism, 683
    of a pyramid, 683
    of three-dimensional
      figures, 699

C

center of a circle, 189

chance experiment, 738

circle, 189–190
    area of, 222
    center of, 189
    circumference of, 190, 196
    diameter of, 190
    radius of, 189–190

circumference, 190, 196
    distinguishing from area, 241

commission, 318

complementary angles, 622

constant of proportionality, 90
    from equations, 104, 111, 132, 172
    from graphs, 152, 157, 165, 172
    from tables, 90, 97, 124, 172
    in equation for
      circumference, 196

coordinate plane, 145
    distance on, 396

corresponding
    angle, 12–13
    distances, 28
    parts, 12–13
    points, 12–13
    segment, 12–13
    sides, 84

cross section, 683

cubic unit, 691, 712

D

deposit, 374

diameter, 190

discount, 318

distance
    on a coordinate plane, 396
    on a number line, 389

distributive property, 270
    in equivalent expressions, 523,
      584, 603

double number line, 291

drawing triangles, 668, 674–675

E

equally likely, 738

equations
    of proportional relationships,
      104, 111, 117, 132
    solving two-step, 502–503,
      509–510, 517, 523

equivalent expressions, 478, 523, 580, 584, 592, 598, 603

equivalent ratios, 84, 90

event, 738

expand, 584

experiment, 731

expressions
    algebraic, 432, 580, 592, 603
    numerical, 432

F

factor (an expression), 584

G

gratuity, *See tip.*

H

hanger diagrams, 502–503, 509–510

I

inequalities, 544
    graphing, 559
    modeling with, 573
    solving, 551–559

interest, 318

interquartile range (IQR), 832

L

like terms
    combining, 592, 598, 603

likelihood, 738

likely, 738

long division, 276